Creation and Detection of the Excited State

Volume 2

Volume 2 in a Series of Monographs

Creation and Detection of the Excited State

Edited by WILLIAM R. WARE

Department of Chemistry
University of Western Ontario
London, Ontario
Canada

VOLUME 2

MARCEL DEKKER, INC., New York 1974

MARCEL DEKKER, INC.
270 Madison Avenue, New York, New York 10016

LIBRARY OF CONGRESS CATALOG CARD NUMBER 76-134785
ISBN 0-8247-6113-8

PRINTED IN THE UNITED STATES OF AMERICA

PREFACE

This volume of the Creation and Detection of the Excited State is intended to adhere to the philosophy and goals adopted by Angelo A. Lamola, who initiated this series and edited the two parts of Volume 1.

This series is intended as a source of information on experimental techniques applicable to the study of all aspects of the formation and behavior of excited molecules. Contributions provide experimental details well beyond what is normally included in a paper or even in an article for a scientific instrument journal. They contain not only a discussion of underlying theory but also examples illustrating the technique, experimental artifacts, corrections, calibration methods, and the scope and limitations of the technique. Chapters aim at providing someone new to a particular area with sufficient information to permit establishing a particular technique in his own laboratory, given hardware and the normal skills of an experimentalist. The contributions are looked upon not as a comprehensive review but as a detailed account of a technique or area. To a considerable extent, the authors draw upon their own experience and instrumentation and take many examples from their own work. However, major contributions to a technique from other laboratories are generally described, sometimes in considerable detail.

The present volume includes two chapters on what might be termed resonance methods. In the first, M. A. El-Sayed and J. Olmsted describe the very elegant and powerful phosphorescence microwave double-resonance technique, which is proving to be a rich source of information on both the mechanism of intersystem crossing and the nature of triplet states. The second chapter by J. R. Bolton and J. T. Warden deals with the marriage of optical and

ESR techniques. In this chapter Bolton and Warden discuss various aspects of optical production of radicals in the ESR cavity and their detection by ESR. This chapter emphasizes time-resolved ESR-optical spectroscopy and ends with a description of the novel simultaneous optical-ESR measurement of transient spectra.

Two chapters on modern laser techniques complete this volume. In Chapter Three, M. Malley describes picosecond laser techniques that have permitted experimentalists to leap from the nanosecond to the picosecond time range in a few years. In Chapter Four, A. Dienes, C. V. Shank, and A. M. Trozzolo describe various aspects of dye-laser technology in sufficient depth to permit the reader to modify or improve his own dye laser. In addition, they discuss several interesting applications.

W. R. Ware

University of Western Ontario

CONTRIBUTORS TO VOLUME 2

JAMES R. BOLTON, Department of Chemistry, University of Western
    Ontario, London, Ontario, Canada

ANDREW DIENES, Bell Laboratories, Holmdel, New Jersey[*]

M. A. EL-SAYED, Department of Chemistry, University of California,
    Los Angeles, California

MICHAEL M. MALLEY, Chemistry Department, San Diego State University,
    San Diego, California

JOHN OLMSTED, III, Department of Chemistry, University of
    California, Los Angeles, California[†]

CHARLES V. SHANK, Bell Laboratories, Holmdel, New Jersey

ANTHONY M. TROZZOLO, Bell Laboratories, Murray Hill, New Jersey

JOSEPH T. WARDEN, Department of Chemistry, University of Western
    Ontario, London, Ontario, Canada[‡]

[*] Present address: Department of Electrical Engineering,
University of California, Davis, California

[†] Present address: American University of Beirut, Beirut,
Lebanon

[‡] Present address: Laboratory of Chemical Biodynamics,
University of California, Berekeley, California

CONTENTS

Andrew Dienes, Charles V. Shank, and Anthony M. Trozzolo

CHAPTER 1

EXPERIMENTAL METHODS IN
PHOSPHORESCENCE-MICROWAVE DOUBLE RESONANCE

John Olmsted, III and M. A. El-Sayed

Department of Chemistry
University of California
Los Angeles, California

I. HISTORICAL DEVELOPMENTS

The effect of microwave radiation in resonance with the zero-
field (zf) transitions of the lowest triplet state of aromatic
phosphorescent molecules in zf and at low temperatures has been observed, developed, and used in different research areas during the
past few years. A number of papers have appeared, and each has extended the field to new areas or demonstrated new effects. As in
any new field, the amount of new information per paper has been high
at the beginning, a fact that has contributed to the rapid expansion
and development of the field, in spite of the relatively small number of papers published. Only a few groups are presently active in
this field in the United States; there is also one group at Leiden
and one at Stuttgart.

Historically, this field can be related to the optical detection
of magnetic resonance transitions of simple atoms and ions. Research
in optical detection of radio-frequency transitions was first performed in 1925 on gases by Fermi and Rasetti [1] as well as by Breit
and Ellet [2], who observed a change in the polarization of resonance
fluorescence upon a change in frequency of an applied alternating
magnetic field. The first determination of magnetic transitions in
solids was not made until 1959 [3-5]. For a review of the field,
the reader is referred to the book by Bernhein [6].

In 1967, optical detection of ESR transitions of the triplet
state was accomplished [7-9]. This was done at 4.2°K, where the
difference in the Boltzmann population of the Zeeman levels of the
lowest triplet state is large enough that microwave radiation can
have a measurable effect. The state of spin alignment, in which the
population of the zf levels is not Boltzmann, was first detected [10]
in 1966 by observing ESR emission lines, rather than absorption lines.
In zf, the state of spin alignment was first detected in 1967 by observing the nonexponential[*] behavior of the phosphorescence decay

_____

[*] The first observation of nonexponential decay was made
earlier but was not attributed to the presence of spin alignment
[A. W. Hornig and J. S. Hyde, Mol. Phys., 6, 33 (1963).

[11]. The first optical observation of zf transitions was made [12] in 1968. The first application of phosphorescence-microwave double resonance (PMDR) techniques to the understanding of the optical spectroscopy of the triplet state was made [13] in 1969. The same year brought the first optical detection of electron-nuclear double resonance [14, 15] (ENDOR). A large number of advances were accomplished in 1970; e.g., the determination of the relative ISC rates using delay techniques [16] and steady-state saturation methods [17]; the optical detection of EEDOR [18]; and the determination of the polarization of the zf magnetic transitions [19]; and the optical detection [20] of level anticrossing in the triplet state and the change of the polarization of the phosphorescence bands originating from two zf levels upon microwave saturation of these two levels [21]. As of 1971, a number of interesting reports on the phosphorescence-microwave double resonance (PMDR) techniques had appeared. Phosphorescence modulation due to the coherent coupling of the triplet spin system to a strong microwave field was detected [22]. The use of PMDR techniques to determine the coherent length of triplet excitons was reported [23].

This chapter examines the experimental methods used in PMDR spectroscopy. It is not merely a review, but rather an exhaustive experimental examination of the factors that determine the signal-to-noise ratio of the different techniques employed. Most of the results presented are new and have not been published elsewhere.

II. PMDR BASIC EQUATIONS

For a polyatomic molecule such as benzene or napthalene, the ground state has all its electrons paired. Upon the absorption of light, one-electron excitation leads to excited singlet states S_1, S_2, . . ., S_x, depending on the wavelength of the exciting light. In about 10^{-11} sec, spin-conserved nonradiative processes $S_x \to S_1$, known as internal conversion processes, deactivate the molecules to S_1, the lowest excited singlet state. Fluorescence ($S_1 \to S_0$ emission),

intersystem crossing (ISC) processes (lowest singlet → triplet
manifold nonradiative processes), or nonradiative S_1 → S_0 processes
might take place. Those molecules that undergo an S_1 → T_x ISC process find themselves in the triplet state with their two unpaired
spins having a specific direction i in the molecular framework; i.e.,
they cross to the zf τ_i level. If the temperature is low enough that
the spin-lattice relaxation (SLR) between the zf levels is slow or
absent, a fast internal conversion from the τ_i zf level of T_x to
the τ_i zf level of T_1, the lowest triplet state, takes place. The
τ_i zf level of T_1 is populated and decays with rate constants that
can be different from those for the other two zf levels. If the SLR
processes are among the zf levels of the lowest triplet state, then
the population ratios of the zf levels of T_1 depend on the mode of
preparation of T_1, as well as on the constants k for the decay and
K for the ISC process. These different situations are discussed
below, assuming that the SLR is absent.

its steady state. Using the steady-state approximation, the relative population of the different zf levels of T_1, i.e., the degree
of the spin alignment in the lowest triplet state, is given by [24,
25]

$$\frac{n_i}{n_j} = \frac{K_i k_j}{K_j k_i} \tag{1}$$

Two non-steady-state situations could occur. In one case,
pulsed excitation is used, e.g., flashlamps are used whose flash
duration is short compared with the decay times of the zf levels.
In this case, the population ratio is given by [26]

$$\frac{n_i}{n_j} = \frac{K_i}{K_j} \tag{2}$$

The other extreme occurs when the excitation is cut off after the
system reaches a steady state. In this case the population ratio
is given by [16]

$$\left(\frac{n_i}{n_j}\right)_t = \frac{n_i^o \exp(-k_i t)}{n_j^o \exp(-k_j t)} = \frac{K_i k_j}{K_j k_i} \exp[(-k_i - k_j)t]. \ . \ . \tag{3}$$

Spin-lattice relaxation would tend to equilibrate or equalize the population of the zf levels; thus, by definition, it would tend to "wash out" or decrease the degree of spin alignment.

The above three equations are used in three different methods used to determine the ratios of the ISC rates. Equation (1) is used in the cw saturation method [17], Eq. (2) is used in the pulsed excitation method [26], and Eq. (3) is used in the delayed method [16]. In all these methods, the population ratio is determined from the intensity changes when a resonance microwave is used with sufficient power to saturate (equalize) the populations of the two zf levels τ_i and τ_j. Another method was proposed [27] in which the ratios can be obtained by rapidly sweeping a microwave of sufficient power to invert the population of the levels.

PMDR techniques require that the triplet state be formed in a state of spin alignment before the microwave radiation is applied. Upon the application of microwaves of resonance frequencies to the zf transitions, the population of each of the two levels being connected with microwaves is changed. If the emission originates from either level or from both levels with two different decay constants, a change in the phosphorescence intensity is observed when the microwave is applied.

Let us derive the equations for the intensity ratio before and after applying the microwave for each of the three situations discussed above, and for the simple case in which the intensity of the phosphorescence, or that for a band isolated and detected by a spectrometer, originates from a single zf level; e.g., τ_i (i.e., the emission from the τ_j zf level is not detected). If the microwave is pulsed or swept rapidly through resonance in a time short compared with the decay time of the levels, and the microwave saturates the $\tau_i \rightarrow \tau_j$ transition, one can show that

$$\frac{I^{\nu_{ij}}}{I^0} = \frac{k_i^r \left[\frac{(n_i + n_j)}{2} \right]}{k_i^r n_i} = \frac{1}{2}\left(1 + \frac{n_j}{n_i} \right) \tag{4a}$$

where $I^{\nu_{ij}}$ and I^0 are the phosphorescence intensity in the presence and absence of the microwave radiation, respectively.

If, on the other hand, the microwave inverts the population of the two levels, then

$$\frac{I^{\nu_{ij}}}{I^0} = \frac{k_i{}^r n_j}{k_i{}^r n_i} = \frac{n_j}{n_i} \tag{4b}$$

One then substitutes the population ratios given in Eq. (1), (2), or (3) into (4a) or (4b) and obtains the proper equation in accordance with the mode of preparing the state of spin alignment. If the microwave saturates transitions, then one obtains for the different modes of excitation the following equations:

$$\frac{I^{\nu_{ij}}}{I^0} = \frac{1}{2}\left(1 + \frac{K_j k_i}{K_i k_j}\right) \tag{5a}$$

$$\frac{I^{\nu_{ij}}}{I^0} = \frac{1}{2}\left(1 + \frac{K_j}{K_i}\right) \tag{5b}$$

$$\left(\frac{I^{\nu_{ij}}}{I^0}\right)_t = \frac{1}{2}\left[1 + \left(\frac{K_j k_i}{K_i k_j}\right)\exp\left[-(k_j - k_i)t\right]\right] \tag{5c}$$

Equations (5a-c) are applicable to the steady-state pulsed excitation, and delay methods, respectively. If the microwave pulse or sweep inverts the population, the following equations are obtained instead:

$$\frac{I^{\nu_{ij}}}{I^0} = \frac{K_j k_i}{K_i k_j} \tag{6a}$$

$$\frac{I^{\nu_{ij}}}{I^0} = \frac{K_j}{K_i} \tag{6b}$$

$$\left(\frac{I^{\nu_{ij}}}{I^0}\right)_t = \left(\frac{K_j k_i}{K_i k_j}\right)\exp\left[-(k_j - k_i)t\right] \tag{6c}$$

It is obvious that the choice of method to be used to obtain large optical signals (or large $I^{\nu_{ij}}/I^0$) would depend on the rate constants of the levels involved. If $K_j k_i \simeq K_i k_j$, then the steady-state method is not appropriate and one of the other methods would be useful. The choice of one of the other two methods depends greatly on whether the anisotropy of K is greater or smaller than that of k. If it is greater, then the pulsed excitation method is best suited; otherwise, the delay method would give better optical signals.

In the above discussion, the microwave saturates or inverts the level population in a time that is short compared with lifetimes of the levels. There is another situation in which the microwave is swept very slowly so that the population of the levels has a value which is in a steady state not only with respect to UV excitation and decay but also with respect to the absorption of the microwave. This is the case when $I^{\nu_{ij}}$ is the intensity reached while the sample is exposed to resonant microwave radiation for a long time compared with the lifetime of the levels. This is the value used in the originally proposed cw saturation technique [17]. In this case,

$$\frac{I^{\nu_{ij}}}{I^0} = \left(1 + \frac{K_i}{K_j}\right)\left(\frac{k_i}{k_i + k_j}\right) \tag{7}$$

The different experimental techniques used to observe and measure the signal due to the change in the phosphorescence intensity from I^0 to $I^{\nu_{ij}}$ make use of the above situations or combinations of them. The experimental examination of these methods will constitute the rest of the chapter. Before discussing these techniques, it is important to point out that observation of these optical signals is now being used in three main research areas: magnetic resonance, optical spectroscopy, and the field of understanding nonradiative processes. This is accomplished by active studies aimed at determining the energies and properties of transitions; determining the zf origin of the emission of different vibronic bands in the phosphorescence spectrum, and also the assignment of the spatial symmetry of triplet states;

and determining the relative ISC rates and thus the mechanisms of
nonradiative processes involving changes in spin quantum numbers.

The following references are recommended:

1. For optical pumping of the Zeeman transitions of atoms and
ions, see Ref. [6].

2. For the phenomena of spin alignment (production of the
triplet state with its zf levels having non-Boltzmann population),
see Ref. [28].

3. For a recent review of the principals involved in the phosphorescence-microwave multiple resonance techniques, see Refs. [24]
and [29]. For a detailed examination of the methods used in determining the relative rates of ISC processes, see Ref. [30].

4. For an examination of the crucial experimental parameters
of optical detection of magnetic transitions in the triplet state,
see Ref. [31].

5. In this work, coaxial lines [32,34] and coils as well as
microwave horn antenna [42] are used for transmitting the microwave
radiation to the sample.

6. For detailed experimental techniques and optical determination of the polarization of magnetic transitions, see the original
work, Ref. [35].

III. EXPERIMENTAL TECHNIQUES

A. General Discussion of Experiments

There are several methods by which microwave resonances can be
detected and the microwave spectrum measured. The methods utilized
in our laboratory, in order of increasing complexity of the attten-
dant electronics are: (a) continuous wave (cw) operation; (b) modulation of the microwaves with lock-in detection at the modulation
frequency (lock-in method); (c) cutoff of the exciting light beam
followed by a sweep of a microwave frequency region after a specified

delay time (delay method); and (d) sweep of a microwave frequency region after subjecting the sample to a short-lived light excitation pulse (pulse method). Each of these methods incorporates certain advantages; which one is most advantageous for a given molecular system will depend on the rate constants for population ($S_1 \rightsquigarrow T_1$ intersystem crossing) and depopulation ($T_1 \rightsquigarrow S_0$ radiationless or phosphorescent decay) of the individual triplet sublevels.

Optical detection of microwave transitions involves irradiating the sample with light of energy sufficient to populate the triplet sublevels, exposing the sample to microwave radiation, and observing changes in the phosphorescence emission intensity as the microwave frequency is scanned. Population of the triplet sublevels is most commonly achieved by optical pumping of the molecule to its first excited singlet state, followed by intramolecular intersystem crossing to the triplet manifolds; but any other pathways leading to the triplet state (singlet-triplet absorption, host-guest energy transfer, etc.) can equally well suffice, and microwave resonance studies can yield important information about such pathways.

In order for optical detection of microwave resonance to be feasible for a given molecular system, three important conditions must be satisfied: (a) the triplet state in question must display luminescence; (b) some mechanism must exist for creating a significant population of molecules in the triplet state with a population imbalance between the individual sublevels; (c) this population imbalance, once created, must persist for a time sufficient to allow interrogation of the system by microwaves.

The first condition, though by no means universally satisfied, is not a serious limitation, since many molecules, both organic and and inorganic, display phosphorescence when in glass or solid solution at reduced temperature (77°K or lower). The second of these conditions is generally readily satisfied, inasmuch as intersystem crossing processes not only exist with significant yields for many molecular systems but also are subject to spin state symmetry selection rules which cause them preferentially to populate specific sublevels in the triplet manifold.

Satisfaction of the third condition depends on the rates of the spin-lattice relaxation processes by which the spin sublevels inter-convert; these rates are highly temperature dependent and depend also on the environment within which the molecular system is placed. In order to maintain a steady state triplet sublevel population im-balance, the rates of sublevel interconversion must be slower than the rates at which the sublevels depopulate back to the ground sing-let state. To reduce the spin-lattice relaxation rates sufficiently, temperatures of the order of $5°K$ or lower are required. In some solid matrices, the "freezing" of the spin lattice relaxation is effectively complete at $4.2°K$, whereas in other matrices, it may not be attained even at $1.4°K$. Whether spin-lattice processes can be "frozen" at liquid-helium temperatures may even depend on the solid phase of the crystals. For pyrazine in cyclohexane [36], for example, slow cooling of the solution to below its freezing point prior to im-mersion in liquid helium gives a spin-polarized system, but rapid cooling yields a system in which spin-lattice relaxation rates are too fast even at $1.4°K$ for polarization to occur.

In principle, optical detection of microwave resonances should also be possible, even in the presence of fast spin-lattice relaxa-tion, using a pulsed light source to create an instantaneous triplet sublevel population imbalance and observing the effect of microwaves before the populations have an opportunity to relax. Although such an effect has been observed using ESR techniques at $77°K$ [41], the experimental difficulties involved in doing optical observations of this type at this high temperature have so far prevented the effect from being observed in the optical mode.

The basic instrumentation required for optical detection of microwave resonances is the following: (a) an excitation light source (the higher the brightness, the easier the detection) equipped with lenses and filters sufficient to raditively pump the sample to an electronic excited state from which the triplet state is efficiently populated; (b) a variable-frequency microwave generator or sweeper in the 0.2-12 GHz range, together with the hardware required to con-duct microwaves to the sample (connectors, coaxial cables, stainless-

steel coaxial rod); (c) a conventional quartz-tipped liquid helium
Dewar for spectroscopic measurements, modified suitable to conduct
microwaves to the sample and to be pumped on rapidly to reach lower
temperatures (∿1.4°K); (d) detection components of the type commonly
used for time-varying emission spectroscopic measurements, such as a
phototube or photomultiplier tube and an electrometer or oscilloscope.
As in most spectroscopic experiments, the level of sophistication of
the apparatus can range from quite simple to extremely complex, depending on the specific requirements of the experiment at hand. Some
of the possibilities for apparatus variations are described in the
subsections which follow.

A schematic sketch of a typical experimental setup is shown in
Fig. 1. The sample is inserted into a copper coil slow-wave helix
structure which is attached to a stainless-steel coaxial waveguide.
This assembly is immersed in liquid helium in a double-jacketed,
bubble-free, quartz-window Dewar assembly. Exciting light from a UV

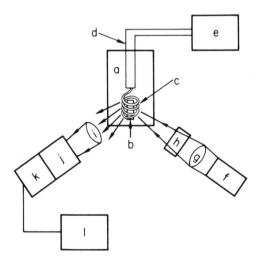

FIG. 1. Schematic diagram of components used in optical
detection of microwave transitions: (a) liquid helium cryostat;
(b) sample; (c) copper coil helix; (d) coaxial waveguide; (e) microwave generator; (f) light source; (g) quartz lens; (h) filters; (i)
collecting lens; (j) monochromator; (k) photomultiplier; (l) electrometer and recorder.

light source suitable for excitation of the phosphorescence emission
of the sample is focused on the sample through suitable filters,
which block unwanted light that would otherwise increase detection
noise. The emitted light from the sample is collected by a lens
system (optional), passed through a filter or monochromator, and detected by a phototube, whose output is amplified and displayed using
suitable electronics.

The experiment then consists of changing the frequency output
of the microwave generator, whereupon, at the frequency corresponding
to a transition between the magnetic sublevels of the sample, an intensity change in the phosphorescence emission will be observed, provided that the microwave power and the population difference between
the levels are both sufficiently large. This basic scheme remains
the same for the various modifications described in the remainder of
this section--modifications which represent ways of enhancing the observed signal, either by increasing the selectivity of the detection
network or by increasing the population difference at the instant of
passage through the microwave resonance.

In a few special cases, the intensity change upon passage through
a microwave transition may be observed with the naked eye. For pyrimidine in benzene, for example, the total emission intensity increases
by about 20% upon passage through the 0.923 GHz (2E) transition (this
change is, however, extremely sensitive to the microwave sweep rate;
see below), and the resulting flashes of light can be readily seen
with the naked eye. Such large changes are unusual; intensity variations of the order of a few per cent are the norm.

The apparatus necessary for doing microwave-optical double resonance experiments is conveniently subdivided into four basic sets of
components as noted above: an excitation lamp, a microwave generator,
a liquid helium Dewar, and a luminescence detector. Certain special
considerations are involved in selecting each of these components.

The essential criterion for the excitation lamp is production
of a photon flux of high density, focused at a spot, and in the energy
range where sample absorption is high. For most applications, a
small-arc (high brightness) mercury light source of about 100 W or

greater output, when used with appropriate filters, is satisfactory.
For work where variation of the excitation wavelength is desirable,
a better choice is an intense continuous light source such as a
1000-5000 W xenon lamp coupled with a small, low f-number monochromator. Because many of the molecular systems that are studied absorb radiation only in the wavelength region below 320 nm, focusing
lenses, filter solution cells, and optical windows through which the
light beam must pass should be of a high-grade quartz such as Suprasil.

Filters for the exciting light beam must be selected to transmit
maximum light within the absorption region of the sample and minimum
light in the wavelength range where the sample emits. The filter
combination chosen will therefore depend largely on the particular
molecule under study. Corning or Jena glass filters combined with
filter solutions such as those described by Kasha or Calvert and
Pitts [37] are the usual choice.

Lamp stability is an important factor, particularly when low-
level signals are encountered which require time-averaging techniques
to be made quantitative. Because the triplet state lifetimes are
frequently in the 10-100 msec range, dc operated lamps are preferable
to ac; otherwise, true steady-state conditions may not be attained.
A low-ripple dc power supply with high long-term stability is thus
called for.

The zero-field microwave transition frequencies for different
molecules encompass the range from below 0.2 GHz to values exceeding
12 GHz. Since individual microwave signal generators cover only a
fraction of this range, one requires a number of such generator
units, each capable of producing signals over a portion of the region.

Conduction of microwave power to the sample also poses certain
problems. The sample itself can be enclosed either in a copper coil
helix structure [see Fig. 2(b)] or in a microwave cavity; the former
is relatively easy to prepare (in fact, a satisfactory coil can be
prepared by wrapping copper wire tightly around a cylindrical tube
such as a large nail or a round pencil, then pulling the coil apart
until the open spaces equal the wire diameter), and will conduct

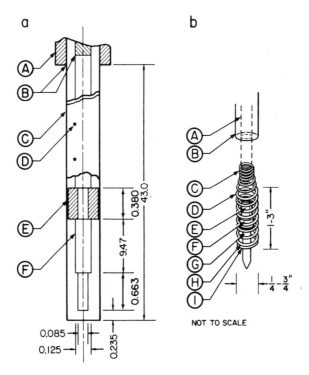

FIG. 2. (a) Details of construction of stainless-steel coaxial
microwave conductor: (A) UG-29A/U coax coupler (shown in cut-away
view); (B) solder joints; (C) 0.316 in. o.d., 0.273 in. i.d. nonmagnetic stainless-steel tubing; (D) 1/32 in. holes, several sets spaced
along tubing; (E) split-ring Teflon spacer (shown in cut-away view);
(F) 0.125 in. diam. nonmagnetic stainless-steel rod.

        (b) Details of helix structure and mounting: (A) cen-
tral conductor; (B) outer sheath; (C) tightly wound portion of helix,
fits snugly on central conductor; (D) uniformly wound open portion
of helix, spacing between turns equal to wire diameter; (E) single
crystal sample, supported on (F) cork support and spacer; (G) Pyrex
or quartz ampule for liquid or polycrystalline sample; (H) helix ter-
mination, soldered closed; (I) ampule supported by electrical tape.

microwaves over essentially a continuous frequency range but is

subject to rather high power losses. The latter, on the other hand,

exhibits low power losses but only conducts at specific resonance

frequencies and cannot readily be tuned over much of a frequency

range. Copper helices have been extensively utilized in our laboratories, whereas van der Waals' group has utilized both helix and
cavity structures. Details of helix and cavity design can be found
in several sources [32-34].

Conduction of the microwaves from outside the Dewar assembly to
the sample is generally accomplished by a stainless-steel (to cut
down heat leaks along the coaxial line) tube with a stainless-steel
rod coaxially mounted in its center [see Fig. 2(a)]. Figure 2(a)
shows a schematic diagram of the coaxial line. Connections from this
line to the signal generator are usually via coaxial cable, although
power losses along such cables become excessive at frequencies above
about 9 GHz, and waveguide structures are probably to be preferred.

In order to obtain maximum signal intensities, sufficient power
should be delivered to the sample to saturate the two sublevels (i.e.,
to equalize the sublevel populations). Furthermore, some of the information that can be extracted from microwave experiments depends for
its validity on saturation of the transition under study. The microwave power actually reaching the sample is thus quite critical. Unfortunately, because of cable losses, imperfect coil coupling to the
coaxial rod, and especially sample reflectance and power variations
along the coil, it is extremely difficult to determine the amount of
microwave power actually being delivered to the sample. On the other
hand, whether saturation of a transition is being accomplished can
be fairly readily determined. The technique is to introduce a fixed
attenuation factor into the microwave output (some microwave generators offer dial control of power output; for others, insertion of a
fixed-dB attenuator plug at the microwave output is the best procedure)
and observe the change, if any, in microwave resonance signal. Only
if the transition is saturated at both power settings will the peak
of the microwave spectrum at the maximum remain the same. A broadening of the spectrum at the higher power level indicates that saturation was not complete at the lower level.

If a transition is found not to be saturated even at maximum
power output of the generator, several ways of increasing power input
to the sample are possible: reposition the sample in the helix; im-

prove the microwave coupling by using shorter cables and/or more
carefully designed helices; use a traveling-wave amplifier to increase
the microwave power output; or replace the helix with a microwave
cavity tuned to the resonance frequency.

B. Frequency Measurements

Inasmuch as frequency counters for the GHz range are not readily
available, accurate determination of microwave transition energies
is most commonly accomplished by frequency mixing techniques. Using
a transfer oscillator such as the Hewlett Packard (HP) Model 540B, the
microwave frequency which is to be determined is mixed with a variable
frequency ν_2 in the 100-200 MHz range. If the variable frequency is
tuned to a point at which the two frequencies beat with one another
and then fine tuned to a null, the relationship

$$\nu_1 = n\nu_2 \tag{8}$$

is satisfied, where n is an integral. Repeating the nulling procedure at the next higher beat frequency yields

$$\nu_1 = (n + 1)\nu_3 \tag{9}$$

In the 100 MHz range, frequencies can be determined to seven-place
accuracy with an electronic counter-frequency converter combination
(such as an HF 5245L electronic counter with a 5253B frequency converter plug-in unit). Thus, knowing ν_2 and ν_3, ν_1 can be readily
determined. Usually the frequency value is already known to within
5% from the microwave sweeper setting; in this case, a single measurement of a subharmonic frequency is sufficient.

For transitions which can be readily observed under unswept cw
conditions, frequency determination becomes simply a matter of manually tuning the microwave frequency until the transition maximum is
reached, and then feeding the microwave output to the transfer oscillator and performing the above described operations to determine this
frequency.

When the transition is weak, highly structured, or must be rapidly swept to develop significant signal, more elaborate techniques
must be employed. The signal display of such transition is normally
in the form of an x-y plot of signal intensity versus time. To determine the frequency of the transition, one must measure accurately
the starting and stopping frequencies and also determine the linearity
of the microwave frequency versus ramp voltage curve. Further, it is
necessary to know the sweep time accurately. These latter factors
generally require extensive and time-consuming calibration procedures.

A substantially simplified procedure may be used if the microwave
sweeper has a reasonably accurate ΔF sweep mode with good sweep
linearity within the ΔF range (the HP 8690B meets these requirements).
In such a case, one can determine the center frequency accurately
using a transfer oscillator, and then do two determinations of the
transition position, using identical center frequency, sweep rate,
and display conditions but altering the frequency range by a factor
of 2. Under these conditions, one can show that the frequency to be
determined (ν_u) is given by the expression

$$\nu_u = \nu_c + \Delta F \left(\frac{D_1 - D_2}{2D_2 - D_1} \right) \tag{10}$$

where ν_c is the center frequency, ΔF is the smaller of the two frequency sweep intervals, D_1 is the displacement of the transition from
the starting point of the sweep for the smaller ΔF, and D_2 is the
displacement for the larger ΔF. Use of this technique on a transition
of known frequency using the 8690B sweeper gave an apparent transition
frequency that was within 0.2% of the known values; the deviation is
less than the reported linearity limits (0.5%) for this sweeper.

Selection of the best liquid helium Dewar structure for a given
microwave double resonance system will usually depend to a certain
extent on the optical components with which the Dewar is to be used.
Certain requirements on the cryogenic apparatus are common to all,
however. Although optimization of all factors would require custom
design and construction, standard commercially available liquid helium cryostats can be readily modified to meet these requirements.

In general, to admit exciting light of sufficiently high energy, Dewars with quartz tips or windows are required. Secondly, to allow attainment of temperatures in the 1.5°K range (desirable to ensure spin polarization), a vacuum-tight cap and large mechanical pump capable of reducing the vapor pressure to the millimeter range are needed. A mercury manometer connected to the system provides a convenient temperature measuring device. The stainless-steel coaxial waveguide may be inserted through the vacuum cap using a Kwik-Fit type connection, which provides a vacuum seal but allows for both vertical and rotatory motion of the sample which is mounted in the helix at the end of the waveguide.

Inasmuch as the exciting light impinges on the sample after passing through layers of liquid nitrogen and liquid helium, and the sample emission reaches the detector only after passing through similar layers, it is essential to hold to a minimum disturbances in these liquids which could scatter light. These are primarily four in nature: gen bubbling can be completely suppressed by continuous passage into chanical oscillation.

Bubble formation is primarily a problem in the liquid nitrogen jacket, as the helium tends to boil from its surface. Liquid nitrogen bubling can be completely suppressed by continuous passage into the liquid of a slow stream of helium gas, whereupon all bubbling can be made to take place at the point of introduction of the helium gas stream. A better solution is to design the Dewar such that no liquid nitrogen is present in the portion where the sample is excited and viewed. Light pipes have been used [12] for excitation and/or phosphorescence. Suspended solid particles [primarily ice in the liquid nitrogen and solid nitrogen and oxygen (from air) in the liquid helium] can be excluded by careful flushing of the entire system with clean helium prior to introduction of the cryogenic fluids. Since much liquid nitrogen is contaminated with some ice particles, it is usually necessary also to filter the liquid nitrogen through filter paper or glass wool. To prevent contamination from the atmosphere, the nitrogen jacket is sealed except for a small vent for the escaping nitrogen and helium vapors. For contamination-free transfer of liquid helium, standard cryogenic sources should be consulted.

Thermal current noise arises primarily in the liquid helium and is difficult to eliminate, but if the liquid helium is pumped to below the superfluid (lambda) point, noise from this source is significantly reduced. At the same time, however, vibrations from the pump are generally transmitted to the liquid and to the helix itself and may cause mechanical oscillations that appear as noise on the optical spectrum. Minimization of this source of noise requires as complete as possible isolation of the pump from the cryostat. Shock mountings for the pump, flexible bellows sections in the pumping line, and clamping the line to a solid support between pump and cryostat can all help to achieve such isolation.

The detection system must satisfy two conditions: high sensitivity to the phosphorescence emission of the sample, and a high rejection factor for the scattered exciting light and any fluorescence emissions that may be present. The basic detection schemes that can be employed are two: isolation of the phosphorescence by a filter complementary to the excitation filters (that is, one which transmits well in the relatively longer wavelength region of phosphorescence but absorbs nearly all photons in the fluorescence and excitation light regions), combined with a single-stage phototube (RCA 4409 or similar) which then views the total phosphorescence emission; or collection of the emitted light by a lens system (which usually can be of Pyrex, since phosphorescence emission is generally of wavelengths longer than 320 nm), passage through a monochromator to reject all wavelengths but that of a specific emission band, and amplification by a photomultiplier tube (EMI 6256B or the like). In either case, the phototube output must be fed to an appropriate amplifier and subsequently displayed as a function of time on a strip-chart recorder or oscilloscope screen. The various detection options are discussed in greater detail in succeeding subsections.

C. Continuous Wave Excitation Techniques

The simplest detection scheme utilizes continuous excitation of
the sample phosphorescence by the excitation lamp, continuous exposure
of the sample to microwave power, and continuous monitoring of the
total phosphorescence emission by a filter and one-stage phototube combination. The phototube output is displayed using an electrometer
and strip-chart recorder (for slowly time-varying signals) or with an
oscilloscope (for rapidly varying signals), and one searches for emission intensity fluctuations, in the form of output current (voltage)
variations, as the microwave frequency is manually, mechanically, or
electronically driven through a frequency range which contains the
resonance frequency.

Because detection of microwave resonance in this way involves
observation of a small change in a relatively large signal, bucking
techniques to displace the dc emission level to nearly zero are desirable. One can then increase the detector sensitivity by about an
order of magnitude, thereby magnifying any intensity change about
tenfold. Such bucking does not, unfortunately, improve the signal-
to-noise ratio unless coupled with some integrating technique such as
an increase of the electrometer time constant. What is enhanced is
the magnitude of the displayed intensity change, which can render
small percentage changes in the emission intensity more readily observable.

In an unbucked measurement, the change in intensity that one
observes is superimposed on the intensity itself, giving $\Delta I/I$. In
Fig. 3(a), showing a typical example, $I = 50$ and $\Delta I = 7$. One is
searching, therefore, for a small change in a relatively large signal,
which may not be readily detected. When a bucking potential
sufficient to null the dc intensity is applied, $I_{eff} = 0$, and the
limiting factor for signal detection becomes the noise level of the
signal, since one now observes the change in intensity superimposed
on signal noise, $\Delta I/\text{noise}$. Fig. 3(b) shows this for the same signal
as in Fig. 3(a), with a 50 division bucking potential applied and

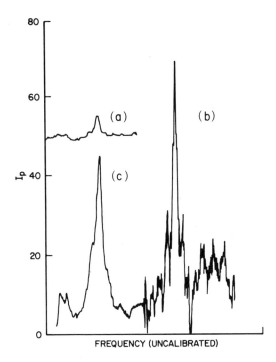

FIG. 3. Effect of applying a bucking potential to improve the sensitivity of an optically detected microwave spectrum (sample is tetrachlorobenzene in durene, 3.67 GHz microwave transition, at 4.2°K). (a) Electrometer on 1 μA scale, 0.1 sec time constant, no bucking current; (b) electrometer on 0.1 μ A scale, 0.5 μ A bucking current, 0.1 sec time constant; (c) electrometer on 0.1 μ A scale, 0.5 μ A bucking current, 1.0 sec time constant.

electrometer sensitivity increased by an order of magnitude. Because of the sensitivity change made possible by the nulling, ΔI = 70 divisions compared with ΔI = 7 divisions in Fig. 3(a). It is clear that fine structure also shows more clearly under these conditions.

Of course, increasing detector sensitivity by an order of magnitude also multiplies the size of the noise display by a factor of 10. This noise can be reduced by increasing the detector time constant, albeit at the cost of somewhat reduced intensity change, possible distortion of the transition line shape, and/or the necessity of sweeping across the transition region at a much slower rate. In Fig. 3(c) the electrometer time constant has been increased by a factor of

10, maintaining other conditions as they were in Fig. 3(b). The loss
in signal magnitude but gain in signal-to-noise ratio is evident.

The signal level ΔI obtained in scanning the microwave resonance
region is found to be highly dependent on the rate at which the microwave frequency is swept through the resonance. Using electrometer-
recorder detection, the sweep rate is stringently limited by the
electrometer time constant and recorder response time in transit
times of the order of 0.1 sec or longer. For molecular systems with
lifetimes of this order of magnitude or longer, such scan rates are
acceptable, but for systems with shorter lifetimes, substantially
faster sweep rates are desirable. This is one of the advantages of
an oscillosope detection system.

The dependence of microwave resonance intensity ΔI on microwave
frequency sweep rate is shown for several different molecular systems
in Fig. 4. The general pattern of variation shows a distinct maximum
in the signal intensity in a certain sweep rate region, falling off
by as much as an order of magnitude or more as the sweep rate is
either increased or decreased from its optimum value. The factors
which affect the position of the optimum point and the extent of decrease as one moves away from optimum conditions are several: the
amount of microwave power applied to the sample (illustrated in the
upper portion of Fig. 4); the width of the microwave transition and
the lifetime of the weakly radiative sublevel being pumped (shown in
the central portion of the figure); and most important, the lifetime
of the strongly radiative level that is being pumped (lower portion
of the figure). The exact sweep rate that optimizes the signal for
a given transition of a certain molecule is determined in a complicated way by the interplay of these factors.

At very fast sweep rates, the microwave transition is being
swept too rapidly, thus decreasing the probability for saturation
(or inversion).* The populations of the two sublevels are not
equalized (or not inverted) during the short time when the system is

---

*Fast-passage adiabatic inversion occurs when the microwave
resonance radiation is swept fast enough and with high power.

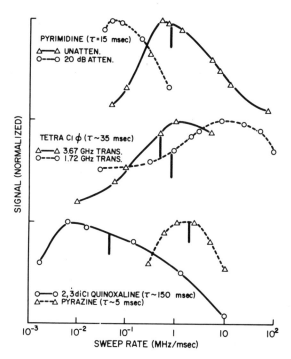

FIG. 4. Variations in microwave resonance intensities with rate of sweep of the microwave frequency for different systems at 4.2°K. Vertical bars indicate sweep rates at which the transition in question is swept in a time equal to the shorter of the two sublevel lifetimes.

on resonance, and less than the optimum signal is developed. From this, it is clear that faster sweep rates can be employed for higher microwave power output. This effect is illustrated by data for pyrimidine (benzene solvent, 4.2°K) in Fig. 4, which shows that the sweep rate at which the signal begins to fall sharply is an order of magnitude slower for 20 dB attenuated microwave power than for unattenuated power (in this case, the maximum signal attained is also some five times smaller for 25 dB attenuated as for unattenuated microwaves).

At sweep rates below the optimum, the decrease in signal intensity may be due to different factors which can operate independently. The first of these is the fact that a slow sweep prevents fast-passage population inversion from occurring, if it were occurring at fast

sweep. The second factor is that a slow sweep represents a situation
in which the emitting system is in a steady state, with the emitting
level having a smaller population change upon microwave irradiation.
This point is described in more detail in the following subsection.
The third factor is that when a transition is swept slowly, transitions due to molecules in different crystal sites may be resolved in
time, such that the signal observed at any particular instant will
be due only to those molecules in a limited fraction of all sites.
When the transition is swept faster, in a time comparable to the lifetime of the strongly radiating level, the different sites are no
longer time-resolved, since molecules in one type of site do not have
time to decay before those in other types of sites are pumped. Thus
the total signal intensity will represent the sum of intensities due
to all sites, which is of course larger than that due to any one site.
Structural features of a microwave resonance can also be caused by
hyperfine interactions; in this case also, sweeping through the entire
transition in a time comparable to the lifetime of the radiating level
results in summation of signals appearing, but with higher signal intensity.

The data illustrated for tetrachlorobenzene (durene solvent,
4.2%K) in Fig. 4 illustrate the effect of two of the factors discussed.
The 1.72 GHz transition couples states having lifetimes of 33 and 37
msec, whereas the 3.67 GHz transition couples the 33 msec lifetime
state with a 700 msec lifetime sublevel. As will be shown below
(see Fig. 8), the former transition is one for which population depletion upon continuous saturation of the transition is not significant, and the decrease in signal with decreasing sweep rate is due
solely to loss of additivity of signals of different frequencies
within the resonance envelope. The intensity levels off at about 50%
of its maximum value at a sweep rate where full structure resolution
takes place. For the latter transition, on the other hand, population
depletion upon continuous saturation is quite extensive; hence the
intensity continues to fall to significantly lower levels as the
sweep rate is decreased and the time during which saturation occurs
is increased.

The role of the lifetime of the more strongly radiating sublevel becomes evident when one looks at the conditions needed to ensure addition of the signals arising at different frequencies within the resonance envelope. These frequencies must be spanned within a time that is comparable to or shorter than the lifetime of the shorter-lived sublevel. Thus, the longer lived the sublevels are, the slower is the sweep rate at which additivity, and hence maximum signal, may be attained. This feature is illustrated for 2,3-dichloroquinoxaline (solvent durene, T = 4.2°K) and pyrazine (solvent cyclohexane, T = 4.2°K) in Fig. 4.

The interplay of these several factors for any particular microwave resonance make it impossible to treat quantitatively the relation between sweep rate and sublevel lifetime, but qualitatively one can predict that given sufficient power (the maximum output, for example, of commercial microwave sweepers), the maximum signal intensity will usually occur near a sweep rate at which the transition is swept in a time comparable to the lifetime of the more strongly radiating sublevel. At such sweep rates, both of the attenuating factors which become important at low sweep rates are minimized, but at the same time, one is sweeping at about the slowest rate consistent with such minimization and thus putting maximum power into the system at the transition frequency. The sweep rates at which the transitions are swept in one lifetime are shown in Fig. 4 by vertical bars.

The shape observed for a microwave transition is highly dependent on the rate of sweeping through the transition relative to the lifetime of the sublevel being pumped. At the low sweep rate limit where the sweep takes many lifetimes to pass through the resonance, the observed line shape will correspond to the microwave spectrum. This is true because the microwave dwells in each small frequency interval for a time sufficient for the system to relax to a steady-state condition. At the fast sweep rate limit, on the other hand, the microwave frequency sweeps through the full transition width in a time which is comparable to or shorter than the sublevel lifetime. Under these conditions, the system is effectively pulsed with a short burst of microwaves of resonant frequency, and the line shape that is observed

represents a rise time upon microwave exposure, followed by a decay
time which is the composite of the return to steady-state populations
of the two sublevels which were pumped. At intermediate sweep rates,
the line shapes are a mixture of these cases and bear no simple physical significance. These features are illustrated for 2,3-dichloro-
quinoxaline (durene solvent, 4.2°K, 2E transition) in Fig. 5. Part
(a) shows the shape observed when the transition is scanned in a time
(10 sec) which is about 100 times as long as the radiating sublevel
lifetime (140 msec). This represents the true microwave spectrum.
Part (b) shows the shape observed when the scan time is only 10 times
the lifetime. Partial resolution of structure is still observed, but
some of the structure, particularly on the long-time side of the resonance maximum, is washed out by the fact that the signal observed at
times after passing through the peak of the resonance is a composite
of the decay of the resonance peak and the microwave spectrum. Part
(c) shows the result of scanning the entire resonance envelope in one
sublevel lifetime. All frequency-varying structure is now obscured;
instead the signal traces out the system's response time to microwaves,
followed by the relaxation times of the pumped sublevels.

These relaxation times depend on the rates of populating the
triplet sublevels from the singlet manifold (intersystem crossing rate
constants) as well as on the rates of depopulation back to the ground
state. However, since the former are typically in the 10^6-10^{10} sec^{-1}
range, the rate-determining step is the depopulation step, with a
rate constant in the 1-10^3 sec^{-1} range. Moreover, the sublevel with
the shorter lifetime is, in general, also the more strongly radiating
level. Thus the decay time that one sees is essentially the lifetime
of the shorter-lived of the two triplet sublevels.

The extent of modification of this lifetime by contributions from
the second sublevel may vary a great deal, depending on the difference
in sublevel lifetimes as well as on the amount of radiation coming
from each of the sublevels. In some cases, where the two sublevel
lifetimes differ markedly but both contribute significantly to the
total observed emission under steady-state conditions, the decay

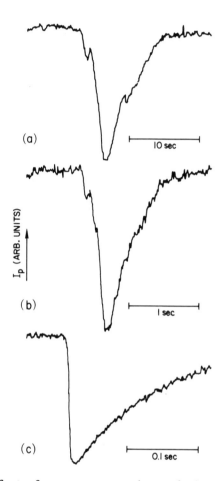

FIG. 5. Effect of sweep rate on observed microwave transition shape (transition is the 2E, 1.055 GHz transition in 2,3-dichloro-quinoxaline, observed at 4.2°K at the nontotally symmetric emission band at 4740 Å). The y axis does not correspond to a frequency axis. (a) Transition swept at a rate sufficiently slow to display accurate line shape. (b) Transition swept at an intermediate rate. (c) Transition swept at a rate sufficiently fast to display accurate rise and decay times.

may be resolvable into an initial, short-lifetime signal decrease
(return of the positively pumped level to its steady-state population)
followed by a longer-lifetime signal increase (return of the nega-
tively pumped level to its steady-state population). Such a situation is shown for pyrimidine in Fig. 6(a).

Although sweep rates for which a transition is scanned during a
sublevel lifetime serve to optimize conditions for observing the
resonance, they clearly do so at the cost of information about the
true shape of the microwave spectrum. This loss of information about
line shape size means that the frequency at which maximum signal intensity is observed may not correspond to the true maximum frequency.
The accuracy of determination of the resonance frequency can be no
greater than the width of the resonance when the sweep rate is such
that signals due to all frequencies within the resonance become additive. If the system rise time is significant compared to the rate of
scanning the resonance, the frequency determination may contain additional error because the signal will not reach its maximum value until
after the resonance frequency has been passed.

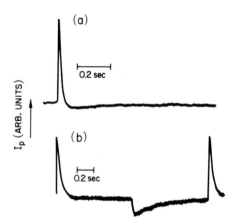

FIG. 6. Transient response of phosphorescence intensity of a
system with greatly different sublevel lifetimes (pyrimidine in
benzene, 2E transition, τ_1 = 15 msec, τ_2 = 330 msec): (a) fast
microwave frequency sweep through the resonance; (b) square-wave
amplitude modulated microwave power, cycled on at the point where
intensity shows a sudden increase and off at the point where it
shows a sudden decrease.

Thus, for accurate microwave resonance frequency determination as well as for accurate microwave resonance spectrum measurements, one needs to ensure that the resonance is being swept slowly enough that the true line shape is traced out. This criterion can be established experimentally in two ways. One is repetitive scanning at slower and slower sweep rates until a rate is reached below which the observed line shape becomes constant. This technique though unambiguous in theory, is difficult to put into practice for transitions whose intensity falls rapidly with decreasing sweep rate, as decreasing signal-to-noise ratio can easily obscure the constancy of the resonance shape. In such cases, an alternative scheme is to sweep through the resonance under identical conditions, first from lower to higher, then from higher to lower frequencies. If the true resonance shape is being traced, the two scans will be mirror images; if not, they will display different shapes (unless, of course, the entire transition is being swept within the sublevel lifetime, in which case the two scans will be identical).

Because frequency and shape determinations must frequently be made under such slow scan conditions that the resonance intensity is far from maximized, a signal accumulation technique [using a computer of averaged transients (CAT)] can be particularly helpful in these measurements. Section G on the use of the CAT describes the instrumentation needed for this technique. Other signal enhancement devices, such as the lock-in detection described in the next section, can also be utilized to advantage.

D. Lock-in Detection Techniques

The use of frequency-selective lock-in detection techniques can improve significantly the sensitivity of microwave resonance detection. In this method, the microwave output of the signal generator is modulated from zero to full amplitude by an externally generated square wave of appropriate low frequency. The same square wave is used as a reference signal for a synchronous amplifier (e.g., Brower 130, PAR HR8, or similar) and the phototube or photomultiplier output

signal is fed to the synchronous detector. Since only the microwave
signal is being modulated at the detection frequency, the detector
sees zero signal except when the molecular system is exposed to microwaves at a resonance frequency. At resonance, the phosphorescence
emission is modulated by the microwaves, and this modulation is seen
by the synchronous amplifier.

The modulation frequencies which can be utilized to advantage in
this technique are limited by the response time of the molecular system, which consists of the rise time for pumping from one magnetic
sublevel to another and the lifetime for decay of the pumped sublevel
after cutoff of microwave power. The former is generally somewhat
faster than the latter, which accordingly sets an upper limit to the
frequency which can be used. If the reciprocal frequency of the
microwave modulation is less than twice the lifetime of the state,
the state population and phosphorescence emission intensity will not
have time to fall back to their steady-state values during the off
cycle of the microwaves. Thus, when the microwave power comes back
on, a smaller intensity change will result and the sensitivity of
the synchronous detector for the microwave resonance will be dimin-
ished.

This feature can be seen most clearly in the lowest portion of
Fig. 7, where, for 50 Hz modulation, the amplitude of the modulated
signal (in this particular instance, the 2E resonance of pyrimidine
in benzene) is substantially lower than under 10 Hz modulation.

On the low-frequency side, two types of behavior must be distin-
guished, depending on whether the sublevel populations are depleted
as one continuously saturates the transition. This in turn depends
on the lifetimes of the sublevels and the amount of nonradiative decay
each contains. The two extreme situations are (a) both levels are
purely radiative with $k_i \ll k_j$ and (b) both levels possess the same
lifetime, with one level purely radiative and the other purely nonradiative.

To illustrate the first type of situation, consider an example
for which the two sublevels are populated by equal rate constants,
but the rates out (purely radiative) are unequal: $k_1 = 10k_2$ ($\tau_1 =$

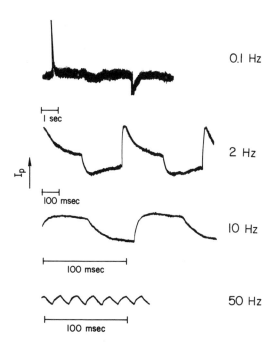

FIG. 7. Oscilloscope tracings of the time variation of phosphorescence intensity when the microwave power, on resonance, is
amplitude modulated at various frequencies. Example of sublevels
of nearly equal lifetimes (2E transition of tetrachlorobenzene in
durene, τ_1 = 33 msec, τ_2 = 37 msec).

$0.1\tau_2$). Then at steady state, in the absence of spin-lattice
relaxation and microwave saturation, $n_2{}^0 = 10n_1{}^0$. The emission
intensity will be given by

$$I^0 = k_1 n_1{}^0 + k_2 n_2{}^0 = 10k_2 n_1{}^0 + 10k_2 n_1{}^0 = 20k_2 n_1{}^0 \qquad (11)$$

A sudden pulse of microwave power at the resonance frequency will
result in equilibration of the sublevel populations without any
level depletion:

$$n_2{}^p = n_1{}^p = \tfrac{1}{2}(n_1{}^0 + n_2{}^0) = 5.5n_1{}^0 \qquad (12)$$

The resulting intensity is

$$I^P = k_1 n_1^P + k_2 n_2^P = 55 k_2 n_1^0 + 5.5 k_2 n_1^0 = 60.5 k_2 n_1^0 \qquad (13)$$

Thus $I^P = 3.02 I^0$. Continuous microwave saturation, on the other hand, will couple the two sublevels so that they decay with a composite rate

$$k_{1,2}^S = \tfrac{1}{2}(k_1 + k_2) \qquad (14)$$

Remembering that the rate of populating the levels remains the same, one can equate the steady-state depopulation rates in the presence and absence of microwave saturation:

$$k_1 n_1^0 + k_2 n_2^0 = \tfrac{1}{2}(k_1 + k_2)(n_1^S + n_2^S) \qquad (15)$$

Substituting appropriately,

$$20 k_2 n_1^0 = 11 k_2 n_1^S \qquad (16)$$

whence $n_1^S = (20/11) n_1^0$, and $n_2^S = (2/11) n_2^0$

The phosphorescence intensity is

$$I^S = k_1 n_1^S + k_2 n_2^S = (200/11) k_2 n_1^0 + (20/11) k_2 n_1^0 = 20 k_2 n_1^0 \qquad (17)$$

Thus $I^S = I^0$, and the intensification of intensity observed when microwave saturation is first attained will, with time, completely disappear.

An experimental example which approaches this extreme is pyrimidine in benzene, for which the behavior of the emission intensity when effectively continuous saturation is achieved (by square-wave amplitude modulation at extremely low frequency) is shown in Fig. 6(b) and the uppermost portion of Fig. 7. When the microwave power

cycles on, a sharp increase of intensity is observed due to the nearly
instantaneous equilibration of population. This intensity increases
then decays at a rate $k^S = \frac{1}{2}(k_1 + k_2) \approx \frac{1}{2}k_1$ as can be verified by
comparison with the signal in Fig. 6(a), representing the same trans-
ition as it appears when the resonance is effectively pulsed by rapid
microwave sweeping. Comparison of time scales shows that $\tau_a = \frac{1}{2}\tau_b$, as
predicted by the above, since in Fig. 6(a) the signal decays at rate
k_1 after the microwaves are off resonance, whereas in Fig. 6(b) the
signal decays at rate $\frac{1}{2}(k_1 + k_2)$ while the microwaves are on resonance.

When the microwave power cycles off, a transient decrease in
intensity is observed, since n_1 falls from its saturated to its
steady-state value at rate k_1, but n_2 rises from its saturated to
steady-state value at the slower rate k_2. This effect can also be
seen in Fig. 6(a), somewhat obscured by being superimposed on the
decay of the resonance signal.

To illustrate the second type of situation, consider an example
where the rates of populating the sublevels differ by a factor of 10,
and the more rapidly populated sublevel decays completely nonradia-
tively at a rate k_2 that is the same as the completely radiative rate
for the other level k_1. Once again, the steady-state populations in
the absence of microwaves will be $n_2^0 = 10n_1^0$. Since only level 1
radiates, the emission intensity will be given by $I^0 = k_1 n_1^0$. Now
an instantaneous microwave pulse which equalizes the populations
without depleting them gives $n_1^P = n_2^P = \frac{1}{2}(n_1^0 + n_2^0) = 5.5n_1^0$.
Hence, $I^P = 5.5I^0$. If the transition is continuously saturated,
$k_1^S = \frac{1}{2}(k_1 + k_2) = k_1$, since the two decay constants are equal.
Thus the decay constants for the two levels are the same whether
or not microwave power is applied, and $I^S = I^P = 5.5I^0$.

This situation is less frequently realized experimentally than
the other, but is approximately the case for the 2E transition in
tetrachlorobenzene in durene, for which the sublevel lifetimes are
33 and 37 msec and one sublevel radiates to the 0,0 band while the
other radiates to the 0,0 - 234 cm^{-1} vibronic band. The 0,0 band
signal observed for this transition at low modulation frequency is
shown in the upper portion of Fig. 8.

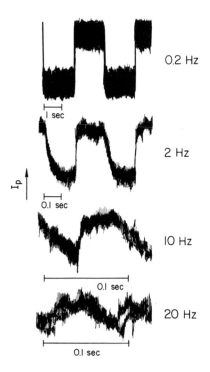

FIG. 8. Oscilloscope tracings of the time variation of phosphorescence intensity when the microwave power, on resonance, is
amplitude modulated at various frequencies. Example of sublevels
with greatly different lifetimes (2E transition of pyrimidine in
benzene, τ_1 = 15 msec, τ_2 = 330 msec).

In general, one expects rate constants for various triplet sublevels that range all the way between the two extremes outlined above.
Therefore, although the decrease in sensitivity of lock-in detection
of higher frequencies can be accurately predicted to depend only on
the lifetime of the more highly radiative state, the behavior at
lower frequencies may be expected to show both sharply peaked (situation a) and plateau type frequency dependence (situation b), as well
as intermediate behavior. Fig. 9 bears this out, showing high-
frequency sensitivity losses for different molecules which correlate
well with lifetimes and also showing one example of sharply peaked

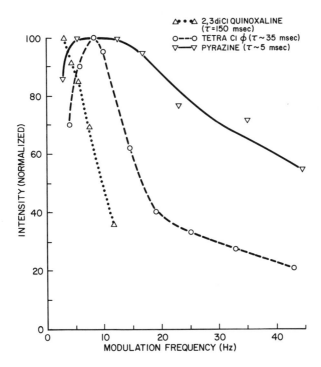

FIG. 9. Variation in microwave transition intensity with modulation frequency using lock-in detection, for systems with various lifetimes.

and one of plateau type low-frequency sensitivity. The frequency to be used for lock-in detection of microwave resonance is therefore best selected after empirical testing of a frequency range extending about an order of magnitude below that frequency which is the reciprocal of the shorter sublevel lifetime.

E. Use of Monochromators

Under most conditions, the signal-to-noise ratio for microwave transitions is substantially better when total emission is viewed

with a simple phototube than when a monochromator-photomultiplier
combination is used to isolate a specific emission band. As an illustration, the pyrimidine signal is shown in Fig. 10 under identical
sample, illumination, and microwave conditions, but in one case with
phototube full-emission detection and in the other with photomultiplier-monochromator detection of the 0,0 band of the phosphorescence
emission. The former offers more than an order of magnitude better
signal-to-noise ratio.

For optical spectroscopic applications, as described in the next
section, use of a monochromator is of course obligatory. In addition,
the photomultiplier-monochromator combination offers potential
advantages under two special sets of conditions.

1. In the presence of emitting impurities or a host which is
luminescent, emission from molecules other than the guest molecule
may produce a very high total emission or strong nearby fluorescence of the molecule under examination; all or some of these may
add substantially to the noise and may even give false microwave
resonances due to host or impurity transitions. In such an instance, the use of a monochromator to isolate emission bands known

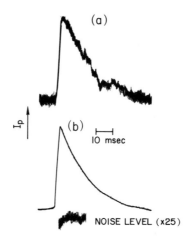

FIG. 10. Comparison of signal-to-noise ratio obtained for microwave resonance transition viewed (a) using monochromator and photomultiplier, and (b) using filter and phototube (2E transition of
pyrimidine in benzene at 4.2°K).

to originate from the guest molecule is a powerful tool for verification of the origin of the microwave resonances. Whether an actual signal-to-noise ratio can also be realized for such systems depends on the relative intensities of desired and undesired radiation. We have observed the system carbazole (guest)-p-dibromobenzene (host), wherein the host shows substantial phosphorescence. We find that the signal-to-noise-ratio of the carbazole microwave resonances is approximately the same for total emission as for monochromator viewing; which is better depends on the detection technique employed. In such questionable cases, especially where signals are weak, both configurations should be tried.

2. The second situation wherein a monochromator may improve detection sensitivity is for molecules for which different bands in the spectrum have comparable intensity and originate from different zero-field levels. In these molecules, different bands would show opposite intensity changes upon saturation at a particular microwave transition. Two examples of this behavior are 2,3-dichloroquinoxaline [38] and 1,2,4,5-tetrachlorobenzene [39]. For such molecules it is obvious that, because of opposing contributions, the percent intensity change when observing the total emission will be smaller than that when observing just one emission band.

Upon saturating the 2E transition in 2,3-dichloroquinoxaline, one observes opposite but unequal changes in intensity of the 4681 Å (0-0) band and the 4740 Å (0,0 - 260 cm^{-1}) band, the two prominent origins in the spectrum. The 2E transition can be very readily observed, using the total emission mode. The total emission intensity built on the 4740 Å band is very much greater than that built on the 4681 Å emission. Only in the unusual situation (of which tetrachlorobenzene is an example) where the total intensity change for all emissions of one type exactly counterbalances that for the other does total emission observation fail; and even in such an instance, it may be possible by judicious filtering of one portion of the emission spectrum to allow one type of emission to dominate.

We should emphasize that the foregoing discussion of total emission versus selected emission pertains primarily to the determination

of resonance frequencies and to the measurement of quantities such as
sublevel lifetimes which are not intensity dependent. For measurement
of quantities that require knowledge of the per cent intensity change
(such as triplet sublevel population ratios), the analysis of total
emission intensity changes needed to yield the desired quantities
might become complicated, particularly in systems like the two described above.

F. Delay Techniques

If one cuts off the light beam exciting the sample, each individual triplet sublevel will be depopulated with its own characteristic
rate constant, and these might be different for the three zf levels
if spin-lattice relaxation rates are slow. As a result, the population ratios of the triplet sublevels at some time t after the cutoff
of the exciting light will be different from the population ratios
under steady-state illumination (t = 0) as shown in Eq. (3).

Under circumstances where $n_1(0) \approx n_2(0)$ but $k_1 \neq k_2$, the population ratio $n_1(t)/n_2(t)$ will then be substantially more favorable for
detection of the microwave resonance at some optimum time after light-
beam cutoff. Using delay techniques, better creation of population
differences between levels τ_i and τ_j is obtained if $n_i^0 \gg n_j^0$ but
$k_j \gg k_i$.

Van der Waals was first to point out [16] the powerful capabili-
ties of delay techniques in determining individual sublevel lifetimes
and population ratios. In addition, it is now realized that using a
CAT (see Section G) in conjunction with the technique can convert it
to a very sensitive probe for detection of zero-field transitions
under certain conditions where they would otherwise be very difficult
to detect.

In the van der Waals technique, the time delay between shutter
closure and passage of the microwave frequency through resonance is
varied. As was pointed out in Section C, for the fast microwave sweep
rates that must be used in this method, the decay of the resonance

peak with time is a measure of the lifetime of the shorter lived radiating levels. The decrease in the amplitude of the resonance peak
as the delay time is increased can be used to compute the lifetime of
the long-lived zero-field sublevel from which pumping is occurring.
Finally, with appropriate assumptions about the relationship between
the amplitude of the resonance peak and the sublevel population, one
can extrapolate to t = 0 and find the steady-state population ratios.
From these values, the relative rate constants for populating the triplet sublevels can also be computed.

In delayed microwave experiments, the excitation light is cut off
at some time t_0 by an appropriate mechanical shutter (such as the
Uniblitz shutter made by Vincent Associates, Rochester, New York),
which has a closure time of the order 1 msec or less. The microwave
frequency is then swept through a transition region at some later time
t which is selected to result in favorable sublevel population ratios.
The apparatus used in our laboratory [40] is shown schematically in
Fig. 11. A mechanical shutter is mounted in the exciting light beam
and provided with a small reference light viewed by a phototransistor.
Upon closure of the shutter, actuated by a square wave of very low
frequency (0.01-0.5 Hz), the reference light beam is also blocked off
from the photoconductor, causing a step change in the phototransistor
output at time t = 0. This change is introduced into a delay box and
is used to generate, at some selected time, output pulses of appro-

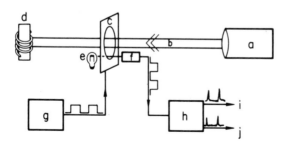

FIG. 11.   Schematic diagram of shutter apparatus for variable-
delay microwave resonance experiments: (a) light source; (b) light
beam; (c) shutter; (d) sample; (e) reference light; (f) photocon-
ductive cell; (g) shutter drive unit (dc power supply plus square-
wave generator); (h) variable-delay pulse generator; (i,j) connectors to microwave sweeper and CAT.

priate sign and amplitude to trigger the microwave sweeper and detector
electronics (for circuitry details, see [40]. Since the populations
of all sublevels are exponentially decaying with lifetimes of the
order of seconds or shorter, electrometer-recorder detection is not
feasible in this technique, as the microwaves must be swept over the
transition in a time which is short compared to the lifetime of the
sublevels. Fast detection with an oscilloscope [16] or CAT [40] is
therefore essential.

The type of signal enhancement that can be obtained using such
delay techniques is illustrated for pyrazine in Fig. 12. This molecule is particularly suited to enhancement in this way, since the
steady-state population ratios of the triplet sublevels are close to
unity, but the emitting sublevel has a lifetime (6 msec) which is
more than 10 times shorter than the lifetimes of the other two sublevels (200 and 600 msec). The enhancement in both signal intensity
(about 4X) and in signal-to-noise ratio (about 11X) is evident in the

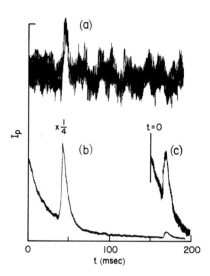

FIG. 12. Enhancement of microwave signal by variable-delay
techniques, D - E and D + E transitions of pyrazine in cyclohexane
(T = 4.2°K, 10.0-10.5 GHz region, τ_1 = 6 msec, τ_2 = 200 msec, τ_3 =
600 msec): (a) microwave sweep with continuous illumination; (b)
microwave sweep after shutter closure at t = 0; (c) microwave sweep
of D + E transition only after shutter closure at t = 0.

figure. In fact, the higher-frequency transition, which is easily observed with the delay technique, cannot be detected at all with single-sweep cw detection and can only be seen with difficulty upon use of multiple-sweep signal accumulation using a CAT.

The effect of varying the delay time for the D - E microwave resonance of 2,3-dichloroquinoxaline is shown in Fig. 13. The decrease in resonance amplitude with the delay time between light cutoff and passage through the resonance is readily seen; if plotted semilogarithmically, the change of this amplitude with time yields a straight line from whose slope the triplet sublevel lifetime is easily obtained.

In the usual case, although the population ratio for the two triplet sublevels may be greatly enhanced with time after shutter closure, the absolute population of each level is attenuated compared to continuous illumination conditions. Thus the value of the population difference becomes small and is frequently very unfavorable for observation of the microwave resonances. For this reason, extension of van der Waals [16] type variable-delay experiments to other than the most favorable cases requires very sensitive detection methods. Because of the short time scales involved, lock-in techniques may not be employed, and one is therefore forced to turn to noise-averaging

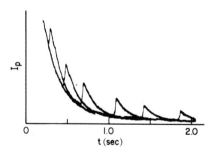

FIG. 13. Decay of microwave transition signal as delay time after shutter closure is increased. Each peak represents the phosphorescence intensity change when the microwave transition was scanned at that particular time after shutter closure (shutter closes at t = 0). Six different observations, with six different delay times, are shown (D - E transition of 2,3-dichloroquinoxaline in durene, 4.2°K, τ_1 = 140 msec, τ_2 = 2700 msec).

methods utilizing signal accumulation over many single sweeps. In this type of application, the CAT is a powerful tool.

G. Computer of Averaged Transients (CAT)

Basically, a computer of average transients (CAT) such as the Varian 1024 or TMC 401C (both of which are used in our laboratory) samples the magnitude of the signal supplied to it in many equal time intervals after a triggering pulse activates it. The magnitude of the signal is first converted [36] to frequency by a voltage-to-frequency converter and the frequency measured as a number of pulses within the time interval being sampled. The number of pulses is stored in a multichannel analyzer. After a single sweep through these channels, the CAT contains a certain number of counts in each channel, the number being proportional to the signal voltage which was being applied to the input of the CAT during the time interval of that particular channel. Thus, in a single sweep, the CAT obtains and displays information that is essentially identical to that provided by an oscilloscope: input voltage as a function of time.

The advantage of the CAT over an oscilloscope comes in extracting low-level signals that are buried in noise, which a normal oscilloscope with its single-pass display could never enhance. The CAT, on the other hand, stores the information of each sweep as counts in the various channels. Thus, if one repeats the triggering pulse and sweeps through the channels again under identical conditions, the counts arising from the new sweep are added to those already residing in each channel (time subinterval). As one repeats this process many times, random noise is gradually averaged out, as it sometimes generates more counts (higher voltage) in some channels (time subinterval), sometimes in others. The signal (voltage increment), on the other hand, always occurs in the same channels and is thereby built up.

When using a CAT in a microwave delay experiment, the phototube or photomultiplier output is fed into the CAT input. The CAT sweep and microwave frequency sweep are simultaneously triggered at a selec-

ted time after shutter closure. Then, as the microwave frequency is
swept through the transition, the CAT samples the phosphorescence
emission intensity change (much as the oscilloscope would). The shutter is cycled open and closed, and with each successive closure the
sweeps are repeated, until a sufficient number of sweeps have accumulated to yield a reasonable microwave resonance signal. The number
of sweeps needed may vary from ten to as high as 300. The accumulated
signal is then printed out by a digital or x-y recorder.

Use of a CAT is not limited to delay type experiments. It can be
used for signal-to-noise enhancement in any application where the variation of the signal with time can be accurately reproduced. Thus it
can also be used in continuous-illumination experiments where microwaves are being swept through a transition. Since most CATs have
provisions for varying the per-channel dwell time between 1 msec or
less and 1 sec, a particularly useful application in cw experiments
lies in accurate line-shape determination of weak resonance transitions. Here, slow microwave sweep rates are coupled with long per-
channel dwell times, so that the variation in phosphorescence emission
with time as the resonance is swept represents the true line shape of
the resonance. Repetitive sweeps through the resonance then allow
the signal to be built up relative to random noise and will bring out
the hyperfine structure. In particular, this may permit the discovery
of weak satellite resonances due to isotope effects, to secondary
crystal sites, or to forbidden, low-intensity hyperfine transitions
(see Fig. 5).

H. Modulation Using the CAT

Amplitude modulation of the microwave power, described above in
the section on lock-in detection, can be combined with signal accumulation using a CAT to give a display of the manner in which the phosphorescence intensity varies with time as the microwave frequency, on
resonance, is cycled on and off. The signal generator microwave output is modulated by an externally generated square wave in the same

manner as described in a previous section. At the same time, the synchronization pulse output of the square wave generator is utilized to
trigger the CAT, and the modulation frequency and CAT dwell time are
adjusted so that slightly more than one full modulation cycle is observed by the CAT. Since the CAT is synchronized with the square-wave
modulation of the microwaves, the display on the CAT represents the
modulation of the phosphorescence emission intensity as a result of
the microwave pumping.

When the modulation frequency is made low compared with the
equilibration times of the system, three different intensities can
be characterized: steady-state intensity in the absence of microwaves
(I^0); steady-state intensity with continuous microwave pumping (I^S);
and instantaneous intensity upon first saturating with microwaves
(I^P).

In the general case where each sublevel T_s is being populated
by a rate constant K_s and depopulated by a total rate constant k_s
which includes a radiative part k^r, it can be readily shown that these
three intensities are given by the equations

$$I^0 = \frac{k_1{}^r K_1}{k_1} + \frac{k_2{}^r K_2}{k_2} \tag{18}$$

$$I^S = \left(\frac{k_1{}^r + k_2{}^r}{k_1 + k_2}\right)\left(K_1 + K_2\right) \tag{19}$$

$$I^P = \frac{1}{2}\left(k_1{}^r + k_2{}^r\right)\left(\frac{K_1}{k_1} + \frac{K_2}{k_2}\right) \tag{20}$$

Although these equations are insufficient in general to allow
solution for the various rate constants, even if the total decay
rates are known from delay type experiments, two limiting cases are
of special interest: (a) $k_1{}^r = k_1$ and $k_2{}^r = k_2$, i.e., both sublevels
decay by purely radiative processes, and (b) $k_1 = k_2 = k$, i.e., both

sublevels decay with the same lifetime. In case (a), the intensity relations become

$$I^0 = K_1 + K_2 \tag{21}$$

$$I^S = K_1 + K_2 \tag{22}$$

$$I^P = \left(\frac{k_1 + k_2}{2}\right)\left(\frac{K_1}{k_1} + \frac{K_2}{k_2}\right) \tag{23}$$

In other words, in this situation the pulsed intensity is different from the steady state, but the two steady-state intensities (saturated and unsaturated) are the same. In case (b), the relations become

$$I^0 = \left(k_1{}^r K_1 + k_2{}^r K_2\right)k^{-1} \tag{24}$$

$$I^S = \left(\frac{k_1{}^r + k_2{}^r}{2k}\right)\left(K_1 + K_2\right) \tag{25}$$

$$I^P = \left(\frac{k_1{}^r + k_2{}^r}{2k}\right)\left(K_1 + K_2\right) \tag{26}$$

In other words, in this situation the pulsed and continuously saturated intensities are the same. (If in addition $k_1{}^r = k_2{}^r$, then $I^0 = I^S = I^P$, and no microwave signal could be observed.

Results of applying this technique to the 1,2,4,5-sym-tetrachlorobenzene molecule (in durene; T = 1.4°K) and monitoring the 0-0 emission band at 3780 Å are shown in Fig. 14. This system is of particular interest in that it has microwave transitions that approximate the two limiting cases described above. Fig. 14(a) shows the phosphorescence intensity variation when the 2E transition, pumping levels with 33- and 37-msec in lifetimes, is modulated. Clearly, $I^P = I^S$ as predicted for near equality of k_1 and k_2. This measurement thus demonstrates that at least one of these two triplet sublevels does not

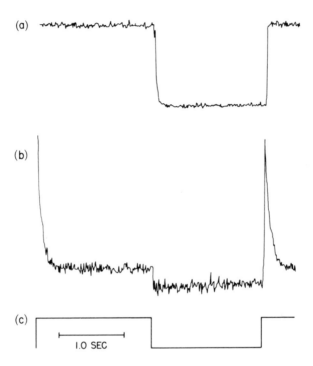

FIG. 14. Phosphorescence intensity changes upon saturation with an amplitude-modulated microwave resonance frequency: (a) 2E transition of tetrachlorobenzene in durene at 1.4°K; (b) D - E transition of tetrachlorobenzene in durene at 1.4°K; (c) modulation pattern of microwave power output (modulation frequency = 0.285 Hz).

contribute significantly to the intensity of this band, as was previously pointed out [39].

Figure 14(b) shows the phosphorescence intensity variation when the D - E transition, pumping levels with very different [39] lifetimes (33 and 700 msec), is equal, indicating that these two sublevels decay nearly entirely by the radiative process. The two observations taken together lead to the conclusion that the uppermost, 37-msec lifetime sublevel, in tetrachlorobenzene does not contribute to the intensity of the 0,0 band, in agreement with previous conclusions [39].

I. AM-PMDR Spectroscopy

The technique of the amplitude modulation of microwaves, described above as a method of improving detection sensitivity for the
microwave resonance spectrum, can also be utilized to advantage in
the measurement of the phosphorescence spectrum. This is the method
of amplitude-modulated phosphorescence-microwave double resonance
(AM-PMDR) spectroscopy [21]. To take such a spectrum, one uses photomultiplier-monochromator detection and first observes a prominent band
in the phosphorescence spectrum. The microwave frequency is adjusted
until it is tuned on a resonance. When the microwaves are set on
resonance, the microwave output is amplitude modulated using an externally generated square-wave of appropriate low frequency (see the
previous section for the criteria dictating the choice of this frequency). The same square-wave frequency is used as reference signal
for synchronous detection; the photomultiplier signal is fed to the
input of the detector, and the detector output is fed to a strip-
chart recorder. When the monochromator is scanned through the wavelength region of phosphorescence emission, an AM-PMDR spectrum results.

Since the synchronous detector responds only to that portion of
the photomultiplier output which is modulated at the reference frequency, the spectrum obtained in this way is composed of only those
emission bands which arise from the two triplet sublevels that are in
resonance. Furthermore, since one level is enhanced and the other
depleted by microwave pumping, some spectral bands will appear as
positive deflections while others will be negative.

A typical AM-PMDR spectrum is compared with a conventional
emission spectrum in Fig. 15. The molecule is sym-tetrachlorobenzene
in a durene matrix at 4.2°K, whose 2E microwave resonance at 1.725
GHz is modulated. The notable comparative features of the spectra,
which are common to AM-PMDR spectra of other molecules as well, are
that the AM-PMDR spectrum appears to contain substantially fewer
bands, and some of the bands appear as negative deflections.

The reduction in number of apparent bands in the AM-PMDR spectrum
is caused by several different factors. To begin with, since the AM-

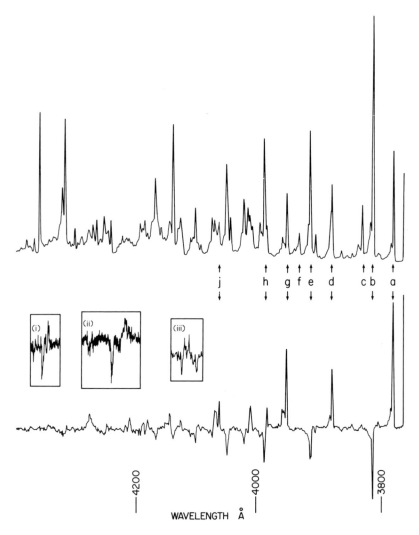

FIG. 15. Conventional (upper portion) and AM-PMDR (lower portion) phosphorescence spectra of tetrachlorobenzene in durene at 4.2°K. Scan speed: 25 Å/min; Spectrometer: Jarrell-Ash 2 m Czerny-Turner; slits: curved, 400 µ; detector time constant (RC): 1 sec; microwave resonance (lower portion only): 2E. Insert (i): Region h of spectrum rescanned with 50 µ slits, 12.5 Å/min speed, 0.1 sec RC. Insert (ii): Region h of spectrum rescanned with 50 µ slits, 5 Å/min speed, 0.1 sec RC. Insert (iii): Region j of spectrum rescanned with 50 µ slits, 50 Å/min speed, 0.1 sec RC.

PMDR spectrum shows only that portion of the emission that is modulated (recall that microwave resonances typically induce only a few per cent intensity change), the signal-to-noise ratio is generally smaller than for the conventional spectrum, causing weaker emission bands to be hidden in noise. Secondly, where closely-spaced bands originate from the different sublevels, their intensities will be additive in the normal spectrum but subtractive in the AM-PMDR spectrum, and with insufficient resolution they disappear entirely. Further, in a molecule of sufficiently high symmetry, different emission bands may originate from different zf levels of the lowest triplet state. Since the microwaves modulate only two sublevels at a time, bands originating from the third sublevel will be unmodulated and therefore absent from the AM-PMDR spectrum, although they would appear (and others disappear) if the spectrum were repeated with the modulation of a different microwave transition. Finally, some of the bands in the conventional spectrum may be due to impurities; since these will not be modulated by the microwaves, they will be absent from the AM-PMDR spectrum.

The spectra shown in Fig. 15 give evidence of several of these effects. A number of the weaker emission bands in the vicinity of the origin are absent in the AM-PMDR spectrum, probably because of an adverse signal-to-noise ratio. In the very rich spectral region between 4000 and 4200 Å, the AM-PMDR spectrum shows evidence of several closely spaced pairs of bands of different origin, and the absence of several moderate-intensity bands from the AM-PMDR spectrum in this region is very likely attributable to cancellation effects. Finally, in the 4300 Å region, two strong bands in the conventional spectrum appear which are missing in the AM-PMDR spectrum. The second of these is the 4358 Å Hg line scattered into the spectrometer and is obviously not subject to modulation. The first represents emission from an impurity, most likely duryl radical, in the tetrachlorobenzene-durene sample.

Because of these features, great care must be taken in selecting spectrometer resolution and scanning rates for AM-PMDR spectroscopy. In general, very good resolution will be required to prevent cancella-

tion effects due to neighboring bands of different zf origin. Moreover, the spectrum will not necessarily provide an indication if insufficient resolution is employed, unlike the conventional spectrum
where broad, partially resolved bands appear. Instead, bands are
likely to be missing at insufficient resolution. This feature is illustrated by (iii) of Fig. 15, showing a portion of the tetrachlorobenzene spectrum, taken on the same spectrometer, with 50 μ slits
rather than 400 μ slits. The negative band, clearly evident at the
end of the region when 50 μ slits are employed, could be very easily
overlooked in the 400 μ slit spectrum.

Unfortunately, going to higher resolution entails loss in intensity which, coupled with the already intrinsically lower signal-to-
noise ratio of the AM-PMDR spectrum, makes observation of all but the
strongest emission bands very difficult. A method which permits ex-
traction of the weaker emission bands from noise involves repetitive
scanning of each spectral region with accumulation of the AM-PMDR
spectrum on a CAT. Application of this technique requires patience
and a spectrometer that will generate a marker pulse at a selected
wavelength. The spectrometer can then be set at a wavelength
somewhat below the marker wavelength (to avoid backlash problems)
and the CAT can be triggered by the marker pulse as the spectrometer
is scanned. A more involved technique must be developed to ensure
that the starting and finishing wavelengths are exactly the same in
each repetition.

In addition to the need for high resolution, accurate AM-PMDR
spectra require the use of extremely slow scan rates. This is partly
due to the fact that synchronous detectors must sample the ac signal
for many cycles to give full response. Since the modulation frequencies used must be low, the time that the monochromator takes to sweep
through an emission band must therefore be of the order of several
seconds. A second consideration is the possibility of time overlap
of positive and negative emission bands. The scan rate must be slow
enough to prevent such time overlap, which would tend to nullify the
signal. This effect is illustrated in (i) and (ii) of Fig. 15
which show a portion of the tetrachlorobenzene AM-PMDR spectrum

scanned at two different rates. The small negative band intermediate between larger positive and negative bands is virtually absent even at the relatively slow 12.5 $\overset{\circ}{\text{A}}$/min rate, but is clearly seen when the rate is reduced to 5 $\overset{\circ}{\text{A}}$/min.

The main advantage offered by AM-PMDR spectroscopy compared to standard phosphorescence spectroscopy lies in the greater ease of making positive assignments for the various emission bands. This becomes increasingly true if AM-PMDR spectra can be obtained and compared for two of the zero-field microwave transitions. In such cases, bands which appear in the conventional spectrum but are absent in both AM-PMDR spectra can be unambiguously assigned to impurities. Furthermore, from the relative intensities of bands in the AM-PMDR spectrum compared to those in the conventional spectrum, vibrations belonging to the same symmetry species can be identified. From Fig. 15, it is evident that the bands marked a, d, and g originate from one sublevel. Since a is the 0-0 band, d and g must also be vibronic levels that are totally symmetric. Bands b and e, on the other hand, originate from the second sublevel and will be of the same symmetry class. Bands c and f, from these spectra alone would be tetrachlorobenzene emissions originating from the third triplet sublevel or might be due to impurities; an AM-PMDR spectrum taken for the D - E resonance reveals them to be due to impurities. Bands d and h are both made up of emissions originating from both sublevels. This is clear for band h from the observation of both positive and negative peaks in the AM-PMDR spectrum. It can be deduced for band d by noting that the relative intensity of this band, compared to a and g, is smaller in the AM-PMDR spectrum than in the conventional spectrum. This feature along with the different shapes of this band in the two spectra indicates its multiple origin. A more powerful method to prove the multiple origin of a vibronic band is to take the polarized AM-PMDR spectrum [21].

Summarizing, AM-PMDR spectroscopy entails a considerable investment of equipment (high-resolution spectrometer, square-wave generator, synchronous detector, possibly a CAT) and time (very slow scan rates, possibly repetitive scans of small regions of the phosphorescence

spectrum), but it can be a powerful tool in identifying spectroscopic
emission bands.

J. Pulsed Excitation with Pulsed (or Rapidly-Swept) Microwaves

A method recently proposed [26] to determine the relative ISC
rates, the phosphorescence mechanisms, and the zf transitions involves
short-pulsed excitation and short-pulsed or rapidly swept microwaves.
If the pulse duration is short compared with the relaxation times of
the zf levels, and if the excitation is much faster than deactivation
of the zf levels, the ratios of the ISC rates to the different zf
levels are given by the ratios of the populations of these levels.
From the changes of phosphorescence intensity emitted from one of
these levels caused by saturating or inverting its population with
that of another level, the relative pumping (ISC) rates of the two
levels can be determined. Furthermore, the resonance transitions can
be determined.

Detection of microwave resonance transitions using pulsed light
and pulsed (or rapidly swept) microwaves requires a suitable flashlamp putting out UV light pulses of millisecond or shorter duration,
some means of subjecting the sample to a pulse of microwave power at
the resonance frequency at a time no more than several milliseconds
after the flash, a photomultiplier detector having a time constant
shorter than a millisecond, and an oscilloscope or CAT with a millisecond time base for signal display.

The light source used in our laboratories was an ILC Model 5L2
flashlamp, 5 mm i.d., possessing a 4-cm arc. The flash was initiated
by a negative pulse, and the output was 30 J. The microwave power was
generated by an HP 8690A microwave sweeper provided with a backward-
wave oscillator plug-in unit, and was transmitted to the sample by
means of a standard coaxial cable, a stainless-steel coaxial microwave conductor, and a slow-wave copper coil helix which contained
the mixed crystal samples. The helix was immersed in liquid helium
in a bubble-free, quartz-tipped liquid helium Dewar which was pumped

with a Kinney mechanical pump to reduce the temperature to 1.4°K.

Emission was viewed through a pair of focusing lenses by a
Jarrell-Ash 0.5 m scanning monochromator equipped with variable slits
which were set in the wide-open position (∿1 mm slit width). An EMI
9502S photomultiplier tube was mounted at the monochromator exit slit,
and its output was observed using a Tektronix 531A oscilloscope with
a type D plug-in unit and a Polaroid camera. The sample was optically
aligned with the monochromator by illuminating it with a 75 W continuous illumination lamp and adjusting positioning to maximize photomultiplier output at the 0-0 emission band of the sample.

A relay box driven by a square-wave generator was used to generate pulses to trigger the oscilloscope sweep and microwave generator
frequency sweep. The same pulses were subjected to variable delay
using a custom-built delay box, and the delayed pulse was used to
trigger the flash unit.

In principle, the light-pulse and microwave-pulse combination
could be generated in any of three ways: (a) a triggered light pulse
followed after a suitable delay by a microwave pulse at the resonant
frequency; (b) a triggered light pulse followed after a suitable delay
by a rapid microwave sweep through the resonant frequency; (c) a
rapid microwave sweep through the resonant frequency with a light
pulse triggered at a time between the start of the sweep and passage
through resonance.

The first of these methods is the most straightforward and would
be most suitable in a system possessing sufficient power output, precisely known resonance frequency, and preferably a microwave cavity
sample holder. With our microwave generator and helix sample holder,
however, it was found that insufficient microwave power was transmitted to the sample in the pulse to achieve saturation. The second
method also was found to be less than satisfactory, in that the microwave frequency sweep circuit possessed a dead time between the activa-
ting pulse and the sweep onset that was of the same order as the
millisecond time scale used in these experiments. The third method
was therefore adopted. The microwave frequency sweep was initiated
at a frequency significantly lower than the resonance frequency (0.1-

0.2 GHz) and the flash activated by a delayed pulse timed to occur a few milliseconds before the microwave sweep passed through the resonance.

Photographs of the oscilloscope tracings obtained for two different microwave transitions using this technique are shown in Fig. 16. The upper portions show the D - E transition of 2,3-dichloroquinoxaline in durene host, at 1.4°K, viewing the nontotally symmetric emission band at 4740 Å (0-0 - 260 cm^{-1}). The lower line on the photograph is the photomultiplier tube output in the absence of the flash. The oscilloscope sweep in this instance was initiated about 2 msec after the flash. The lower portion shows the 2E transition of sym-tetrachlorobenzene in durene host at 1.4°K, viewing the emissions band at 3780 Å (0-0). In this instance, the oscilloscope and flash unit were triggered simultaneously, and the spot at the lower left represents the photomultiplier output in the absence of a flash.

The step change in the phosphorescence intensity in each picture is the result of the population change in the sublevels on passage through the microwave resonance. In 2,3-dichloroquinoxaline, the change is negative because the radiative level is populated with a larger rate constant than the nonradiative. In tetrachlorobenzene, on the other hand, the less radiative level is more rapidly populated, and thus the change upon microwave saturation is positive.

Results for the ratios of populating rate constants for these two systems can be obtained using the equation

$$\frac{K_j}{K_i} = 2\left(\frac{I^\nu}{I}\right) - 1 \tag{27}$$

where I and I^ν are the phosphorescence intensity from zf level i before and after sweeping the microwave across the resonance involving the i and j zf levels. The results for the systems shown in Fig. 16 are in good agreement with values determined by other techniques. For other transitions of these molecules, on the other hand, flash experiments have given ratios that are too low in comparison with those determined otherwise. The lack of agreement is in all proba-

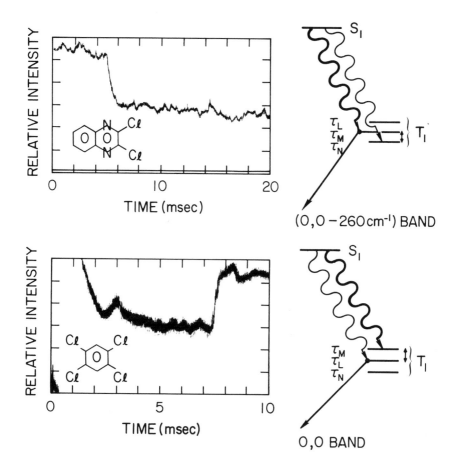

FIG. 16. Determination of the intersystem crossing relative rates for optically detecting zero-field transitions from pulsed excitation followed by a fast sweep of microwave radiation capable of saturating the zero-field transitions involving the radiating level. Top trace is for saturating the $|D| - |E| \tau_M \leftrightarrow \tau_N$ transition of 2,3-dichloroquinoxaline in durene at t = 4.5 msec after pulsed excitation. The (0,0 - 260 cm^{-1}) vibronic band originating from τ_M is monitored. A decrease in intensity upon saturation indicates that the intersystem crossing to the radiating level is faster than that to the other level being saturated. On the other hand, an increase in intensity upon microwave saturation indicates that the intersystem crossing to the radiative level is slower than that to the other zero-field level. This is shown in the bottom trace for 1,2,4,5-tetrachlorobenzene in durene at 1.6°K when the 2E ($\tau_L \leftrightarrow \tau_M$) transition is saturated and the 0,0 band is originating from the τ_L zero-field level (the initial decay observed during the first couple of milliseconds represents the phototube recovery time rather than the flash duration, whose width is shown on the far left of the trace).

bility caused by failure to saturate the microwave transitions in
these other cases and points out one of the greatest obstacles to
carrying out these experiments: difficulty in achieving saturation
of the transition.

This difficulty arises primarily because of the inefficiency of
the slow-wave helix structure in transmitting microwave power to the
sample, compounded in this type of experiment by the rapid sweep
rates required, which reduce the power input to the sample for a
given intensity level. Further, helix efficiency is highly dependent
both on details of helix design and on the microwave frequency being
transmitted. These features make it difficult to determine whether
saturation is being achieved.

Two experimental methods of verifying saturation by microwaves
have been used in this work. The first involves variation of the
rate of the microwave sweep over about one order of magnitude on
either side of the sweep rate used to obtain the results. Since
varying the sweep rate varies the time during which the microwave
frequency is on resonance, such a variation essentially is a change
in the width of the effective microwave "pulse." The power delivered
to the sample is the product of the microwave intensity and the time
on resonance. Since the intensity output is independent of sweep
rate, it is expected that varying the sweep rate changes the microwave resonance power input to the sample. Unless the transition is
being saturated, therefore, the signal magnitude will show a dependence on sweep rate. Such a variation, carried out on the zf transition of tetrachlorobenzene, showed the signal magnitude to be independent of sweep rate at rates up to ten times faster or slower than
that used for the result reported above.

The second method, shown in the top trace of Fig. 16, involves
applying a second microwave pulse at a time somewhat later than the
first one. If the first pulse has saturated the transition, the
second should result in no change in phosphorescence intensity, or
in a slight increase in intensity if the radiative level has had time
to decay significantly in the time between the pulses. If the first
pulse has not saturated the transition, on the other hand, the second

pulse should result in a further intensity change in the same direction as the first.

In the top trace of Fig. 16, the second microwave pulse is applied about 13 msec after the first, by leaving the "blanking" control on the microwave generator off so that microwave power is generated as the sweeper returns to its starting frequency as well as during the sweep. The result is a small positive step, caused by partial decay of the radiative level before the second pulse is applied, which is visible at about t = 18 msec in the photo. (The step is much more clearly evident when signal accumulation on a CAT is used to enhance the signal-to-noise ratio.)

The extension of the optical methods to determine zf transitions, relative ISC rates, and phosphorescence mechanisms to 77°K and higher temperatures is possible, at least in principle. The detection of emission ESR lines has recently been demonstrated [41] at 77°K. Short excitation pulses and high microwave power (to compensate for fast sweeping) are required to excite and detect the system in a time short compared with the SLR times (for systems in which the ISC rate is much faster than the SLR processes at higher temperatures). Complications could arise that would prevent using the simple equations given in the communication [41] if: (a) the excitation or microwave pulse width and delay times are not short as compared with the SLR times; (b) the microwave is not of sufficient power to saturate or invert the population of the levels; (c) the ISC and internal conversion processes are not much faster than the SLR processes.

IV. SUMMARY

In the preceding sections, we have detailed both the basic experiments for optical detection of magnetic resonances and many of the refinements and extensions that are possible. This concluding

section is designed to summarize the usefulness and equipment requirements of these various techniques.

The most basic PMDR experiment requires a liquid helium cryostat, a 100 W small-arc (high brightness) UV light source with filters, a microwave signal generator, and a phototube detection system
including an electrometer and strip-chart recorder. Utilizing continuous illumination, observation of total emission, and microwave
sweep rates of the order of tenths of seconds or longer, it is most
suited to observations on molecular systems where most of the emission originates from one zf level of the triplet state, with lifetimes of tenths of seconds and sublevel steady-state population
ratios that differ significantly from unity.

A simple extension of apparatus to include an oscilloscope
(microsecond rise time is more than adequate, but voltage sensitivity
should be high) will permit higher sensitivity studies to be made on
systems having triplet sublevel lifetimes as short as 1 msec or even
less, provided also that the microwave equipment available has a pro-
vision for sweeping at rates of the order of 100 MHz/msec.

A second relatively straightforward extension, replacement of
total-emission detection by a monochromator-photomultiplier wavelength-
selective detector, offers several advantages. Verification that the
microwave resonances arise from the molecule in question rather than
from impurities or from solvent becomes relatively easy, since optical
spectra and microwave spectra can be correlated by observing microwave spectra in turn on each of a series of the molecular phosphorescence emission bands. Accurate measurement of population ratios requires the use of a monochromator if there is any suspicion of fluorescence emission, solvent or impurity luminescence, or significant
stray-light signals, or even if the phosphorescence originates from
different zf levels for the different vibronic bands, since any of
these features causes the total emission intensities to become a
mixture of additive contributions. Whether the sensitivity of the
detection system to microwave resonances is increased by use of a
monochromator depends on the fraction of total luminescence that
originates from the pumped triplet sublevels.

If a synchronous detector and square wave generator are available as well as a monochromator with a wavelength drive, the powerful
technique of amplitude-modulated phosphorescence-microwave double
resonance spectroscopy (AM-PMDR) becomes feasible. Frequency modulation is very important in detecting zf signals that are otherwise
difficult to detect, because of its ability to sweep rapidly across
resonance. In addition to eliminating impurity emission bands from
the phosphorescence spectrum, amplitude-modulation techniques differ-
entiate between crystal sites for which the microwave resonance frequencies differ. It also is extremely useful in assigning bands in
the optical spectrum to vibrations of different symmetry in those
cases where the emission bands may be originating from different zf
levels. Amplitude modulation can also be used to improve the signal-
to-noise ratio for detection of weak microwave resonances and for
accurate determination of hyperfine structure of the microwave
spectrum.

Probably the most powerful extension of the PMDR technique is
the use of a shutter and variable delay time between its closure and
passage through a microwave resonance. Equally useful, yet not as
fully developed, is the pulsed excitation technique. Unfortunately,
full application of this technique, except to systems showing very
large sublevel population differences, also requires expensive sig-
nal-averaging equipment such as a CAT. With such equipment, one can
determine all of the individual sublevel lifetimes, the extent to
which these are radiative and nonradiative, and the ratios of the
rate constants for populating the sublevels as well.

ACKNOWLEDGMENTS

The authors wish to express their sincere appreciation for the
collaboration of a number of members of their research group at UCLA,
in particular, the contributions of Dr. Chen, Dr. Hall, Dr. Kalman,
Dr. Tinti, and Mr. Leung have been extremely fruitful. We thank

Professor Moomaw and Dr. C. T. Lin for reading the manuscript. The financial support and encouragement of the U.S. Office of Naval Research are also acknowledged with appreciation.

John Olmsted is Visiting Professor from the American University of Beirut, Beirut, Lebanon (1970-1971). This chapter is Contribution No. 2907 of the Department of Chemistry, University of California, Los Angeles.

NOTE ADDED IN PROOF

This article was written in 1971. The field has progressed rapidly and many advances have been made since that time. The reader should consult the literature for new techniques in this field. The two newest techniques that have appeared since writing this review are: (1) the use of optical detection of spin coherence in relaxation studies by C. B. Harris and his group [43,44], and (2) the use of horn antenna as polarized microwave sources to study the structure and orientation of molecules in solids by our group [42].

REFERENCES

1. E. Fermi and F. Rasetti, Nature, 115, 764 (1925); Z. Physik, 33, 246 (1925).

2. G. Breit and A. Ellet, Phys. Rev., 25, 888 (1925).

3. F. Varsangi, D. L. Wood, and A. L. Schawlow, Phys. Rev. Letters, 3, 544 (1959).

4. S. Geschwind, R. Collins, and A. L. Schawlow, Phys. Rev. Letters, 3, 545 (1959).

5. J. Brossel, S. Geschwind, and A. Schawlow, Phys. Rev. Letters, 3, 549 (1959).

6. For a review, see R. A. Bernhein, Optical Pumping: An Introduction, Benjamin, New York, 1965.

7. M. Sharnoff, J. Chem. Phys., 46, 3263 (1967).

8. A. L. Kwiram, Chem. Phys. Letters, 1, 272 (1967).

9. J. Schmidt, I. A. Hesselmann, M. S. de Groot, and J. van der Waals, Chem. Phys. Letters, 1, 434 (1967).

10. M. Schwoerer and R. C. Wolf, in Proc. 14th Collogue Ampere, 1966 (R. Blinc, ed.), North Holland, Amsterdam, 1967, p. 87.

11. M. S. de Groot, I. A. Hesselmann, and J. H. van der Waals, Mol. Phys., 12, 259 (1967).

12. J. Schmidt and J. H. van der Waals, Chem. Phys. Letters, 2, 640 (1968).

13. D. S. Tinti, M. A. El-Sayed, A. H. Maki, and C. B. Harris, Chem. Phys. Letters, 3, 343 (1969).

14. I. Y. Chan, J. Schmidt, and J. H. van der Waals, Chem. Phys. Letters, 4, 269 (1969).

15. C. B. Harris, D. S. Tinti, M. A. El-Sayed, and A. H. Maki, Chem. Phys. Letters, 4, 409 (1969).

16. D. A. Antheunis, J. Schmidt, and J. H. van der Waals, Chem. Phys. Letters, 6, 255 (1970).

17. M. A. El-Sayed, J. Chem. Phys., 52, 6438 (1970).

18. T. S. Kuan, D. S. Tinti, and M. A. El-Sayed, Chem. Phys. Letters, 4, 507 (1970).

19. M. A. El-Sayed and O. F. Kalman, J. Chem. Phys., 52, 4903 (1970).

20. W. S. Veeman and J. H. van der Waals, Chem. Phys. Letters, 7, 65 (1970).

21. M. A. El-Sayed, D. V. Owens, and D. S. Tinti, Chem. Phys. Letters, 6, 395 (1970).

22. K. Schmidt, W. G. Van Dorp, and J. H. van der Waals, Chem. Phys. Letters, 8 (1971).

23. A. H. Francis, and C. B. Harris, Chem. Phys. Letters, 9, 188 (1971).

24. M. A. El-Sayed, Accts. Chem. Res., 4, 23 (1971).

25. M. A. El-Sayed, in Proc. Intern. Conf. Mol. Luminescence, Chicago, 1968 (E. Lim, ed.), Benjamin, New York, 1969, p. 715.

26. M. A. El-Sayed and J. Olmsted, Chem. Phys. Letters, 11, 568 (1971).

27. C. B. Harris, Proc. 5th Mol. Cryst. Symp., Philadelphia, 1970; also J. Chem. Phys., 54, 972 (1971).

28. J. H. van der Waals and M. S. de Groot, in The Triplet State (A. B. Zahlan, ed.), Cambridge Univ. Press, London 1967, p. 101.

29. M. A. El-Sayed, MTP International Review of Science, Spectroscopy (A. D. Buckingham and D. A. Ramsay, eds.), Butterworths, London, 1972, p. 119.

30. M. A. El-Sayed, "Double Resonance Techniques and the Mechanisms of the Intersystem Crossing Processes in Aromatics," in Excited States, Vol. 1 (E. Lim, ed.), Academic, New York, 1974, p. 35.

31. M. Sharnoff, Mol. Cryst., 9, 265 (1969).

32. (a) R. H. Webb, Rev. Sci. Instr., 33, 732 (1962); (b) L. A. Watkins, Topics in Electromagnetic Theory, Wiley, New York, 1958.

33. (a) M. Reisman and R. H. Webb, Rev. Sci. Instr., 38, 1264 (1967); (b) J. R. Pierce, Traveling Wave Tubes, Van Nostrand, Princeton New Jersey.

34. R. E. Collins, Foundations for Microwave Engineers, McGraw-Hill, New York, 1966.

35. O. F. Kalman, and M. A. El-Sayed, J. Chem. Phys., 54, 4414 (1971).

36. L. Hall, Ph.D. Thesis, Univ. California, Los Angeles, 1971.

37. J. G. Calvert and J. N. Pitts, Photochemistry, Wiley, New York, 1966, pp. 729-746.

38. D. S. Tinti and M. A. El-Sayed, J. Chem. Phys., 54, 2529 (1971).

39. C. R. Chen and M. A. El-Sayed, Chem. Phys. Letters, 10, 307 (1971).

40. M. Leung, L. Hall, and M. A. El-Sayed, unpublished results.

41. H. Levanon and S. I. Weissman, J. Am. Chem. Soc., 93, 4309 (1971).

42. M. A. El-Sayed, E. Gossett, and M. Leung, Chem. Phys. Letters, 21, 20 (1973).

43. W. G. Breiland, C. B. Harris, and A. Pines, Phys. Rev. Lett., 30, 158 (1973).

44. C. B. Harris, R. L. Schlupp, and H. Schuch, Phys. Rev. Lett., 30, 1010 (1973).

CHAPTER 2

DETECTION OF TRANSIENT FREE
RADICALS BY ELECTRON SPIN RESONANCE SPECTROSCOPY

James R. Bolton and Joseph T. Warden

Photochemistry Unit
Department of Chemistry
University of Western Ontario
London, Canada

I. INTRODUCTION

Flash photolysis coupled with optical detection of the intermediates has proved to be an invaluable tool in the study of transient
intermediates in photochemical reactions [1]. The widespread application of this valuable technique has resulted in the unraveling of
many photochemical and photophysical reaction mechanisms. Although

a number of physical techniques could be used for the detection of
transient intermediates in flash photolysis systems, optical absorption spectroscopy has remained the principal tool for detecting these
intermediates. However, the recent emergence of flash photolysis
electron spin resonance spectroscopy (FPESR) has provided a new realm
for photochemical and photobiological investigations. Hence this
chapter is more than a review of this rapidly maturing experimental
discipline. It is also a guide to the novice and a projection toward future applications and developments.

The basic principles of electron spin resonance spectroscopy
will not be outlined here. The reader is referred to several recent
books on theory [2], applications [3], and techniques [4,5] for further details.

It is interesting to speculate why ESR has not been used as a
detection technique in flash photolysis systems until recently. The
improvement in sensitivity of the new solid-state ESR spectrometers
is certainly a factor. A computer of averaged transients (CAT) is
essential to the techniques of FPESR: such computers have been on
the market for only a few years. Additional difficulties are posed
by rf shielding requirements and the unique photolytic source intensities and gap sizes required for successful ESR applications. Hence,
although shuttered light sources combined with computer or pen recorder monitoring of the transient ESR signals was popular during the
early 1960s, the successful application of ESR to flash photolysis
studies was not recorded until 1967 [6].

At this point, one might ask why electron spin resonance spectroscopy might be used as a detection technique in preference to
optical absorption spectroscopy. This is a fair question and prompts
a comparison of the advantages and disadvantages of the two detection
systems. Optical absorption detection in a flash photolysis system
is advantageous for the following reasons:

1. Very small concentrations of reactive intermediates can be
detected because of the high sensitivity of the photomultiplier and

in some cases high molar absorptivities. This allows the use of
dilute samples or small-volume sample cuvettes.

    2. Modern technology is extending flash photolysis investiga-
tions into nanosecond and picosecond time domains, thus permitting
an examination of some of the primary photophysical processes.

    3. Where absorption bands are separated and characteristic,
identification of the intermediates may often be made from the optical spectrum.

However, optical detection techniques also possess some limitations. These can be characterized as follows:

    1. Absorption bands from several species may overlap; hence
kinetic analysis of the complicated spectra can be difficult and in-
conclusive.

    2. The general lack of detail in optical absorption bands can
make identification of the observed species uncertain.

    3. Determination of kinetic parameters can be difficult, espec-
ially for transient intermediates, if absolute concentrations must
be determined. Since the intensity of an optical absorption is dependent both on the concentration and on the molar absorptivity (ε)
at the wavelength of interest, unraveling of the kinetic information
requires an accurate value for ε, a factor that is often not available.

    4. Reactive intermediates possessing low molar absorptivities
may be missed.

    5. For successful optical detection, the sample must be opti-
cally clear, hence the presence of scattering particles is to be
avoided. Likewise, the absorbance of the solution must be adjusted
to yield the most efficient use of the monitoring light. Thus high-
absorbance solutions are uncommon.

Although many of these limitations can be circumvented, electron
spin resonance detection can provide information which is not available or difficult to obtain with optical detection. The advantages
of ESR detection in flash photolysis studies can be listed as follows:

1. The technique responds specifically to the free-radical intermediates in the system.

2. Where the free radicals are formed in solutions, the ESR spectra usually contain very sharp lines and overlapping of spectra from different free radicals is usually not a problem.

3. ESR detection is usually not affected by samples which are optically opaque or highly scattering. Hence sample absorbance can be varied over a wide range to produce the optimum signal-to-noise ratio.

4. Each free radical possesses a characteristic g-factor and hyperfine pattern; therefore, identification of transient species by ESR is usually straightforward.

5. If the ESR spectra of two or more free radicals do overlap, the resulting spectra can often be resolved by varying instrumental conditions such as the modulation amplitude and the incident microwave power. Since the linewidth and spin relaxation characteristics differ for each free-radical type, overmodulation or microwave power saturation can be effective in isolating the ESR spectrum of one of the free components.

6. Absolute concentrations of free-radical intermediates can be determined relatively easily by comparison of the area under the absorption spectrum with that of a spin concentration standard [7]. Thus it is possible to determine kinetic rate constants which are other than first order.

Some of the disadvantages of ESR detection are as follows:

1. Transient intermediates with lifetimes less than about 1 μsec may be difficult to observe using ESR detection. This limitation (in the absence of significant spin polarization) is imposed by the spin-lattice relaxation times characteristic of free radicals in solution ($\sim 10^{-6}$ to 10^{-7} sec). However, since some spin polarization is always present, the relaxation criterion is not the primary limitation on the technique in this case.

2. Even if it were possible to detect transient intermediates with lifetimes of less than 1 μsec, the Heisenberg uncertainty principle would preclude identification. Even for a lifetime of 1 μsec, the minimum natural linewidth is approximately 0.2 MHz (or 0.1 G). The resultant loss of resolution due to uncertainty broadening would make identification of transient species extremely difficult.

3. Although presently available ESR spectrometers are capable of detecting about 10^{11} spins per gauss of linewidth, sensitivity considerations are very critical in transient ESR spectroscopy. Since the amplitude of a spin resonance signal is approximately proportional to the inverse square of the peak-to-peak linewidth, transient radicals with wide lines may be difficult to observe with adequate signal-to-noise, even with extensive signal averaging.

4. For free radicals with short lifetimes, the bandwidth of the detection system must be sufficiently broad. However, since the sensitivity is inversely proportional to the square root of the bandwidth (in inverse seconds), an increase in the bandwidth by a factor of 100 will decrease the spectrometer sensitivity by a factor of 10 [4]. Although some sensitivity can be regained by the use of a computer of averaged transients, some loss of sensitivity for short-lived free radicals still remains.

II. COMPONENTS OF THE FPESR APPARATUS

The following sections discuss the characteristics and requirements of various components essential to FPESR. Schematics and block diagrams have been included wherever possible to aid the reader in the assimilation and application of the explanatory material in this chapter.

A. Light Sources and Housings

As with conventional flash photolysis, a number of factors must
be considered in the choice of an optimal photolytic source in FPESR.
Many of these factors are common to both techniques; nevertheless,
the unique geometry inherent in ESR instrumentation places additional
limitations on the type of light source which may be used. Important
factors in the choice of a light source are (a) spectral distribution, (b) light energy per flash, (c) flash duration and profile,
(d) reproducibility, (c) geometrical considerations imposed by the
ESR spectrometer, and (f) compatibility with the ESR spectrometer
and its detection system.

Since photochemical and photobiological investigations often
require actinic sources with wavelengths which may be anywhere in
the UV or visible range, a flash source should have a broad-band
emission. Xenon gas flash tubes (250-1100 nm) are ideal for this
application [31]. If the wavelength range is appropriate, dye and
other tunable or nontunable lasers are often suitable for flash photolysis studies.

Light energy output and flash duration are related, since a decrease in flash duration usually leads to a decrease in light energy
output. As a general rule of thumb, it is advantageous in FPESR for
the photolytic excitation to be completed within the time required
to sample one memory channel in the CAT. For microsecond ESR detection, this requirement necessitates the use of a low-energy ($\sim$2 J
electrical energy) high-repetition-rate flash or a pulsed nitrogen
or Q-switched laser, as the flash duration cannot exceed 1 μsec.
For conventional ESR spectrometers with a response time of about 200
μsec, xenon flashlamps with decay times of 2-20 μsec are suitable.
Here the electrical energy discharged can be about 100 J.

It should be noted that since the minimum detectable number of
spins for an ESR spectrometer is about 10^{11}-10^{13} spins, the flash
excitation source must put out enough photons to generate the requi-
site number of free radicals. Generally, this means that more than

10^{16} photons per flash in the spectral region of interest must be generated. A 100 J xenon flashlamp will deliver about 10^{17}-10^{18} photons to a sample in an ESR cavity.

Quantitation requirements necessitate that the photolytic source be reproducible in energy output and flash duration, although some random fluctuation can be tolerated if a CAT is being used. Aging of flashlamps is a problem, and hence the condition of the tube should be monitored before and after an experiment. A cool stream of dry nitrogen through the housing assists in prolonging lamp life and reproducibility and also removes trace amounts of ozone.

The dimensions of the cross section of an ESR sample tube presented to the actinic flash is about 3 cm x 1 cm at X band. Of this, only the center 1 cm^2 is effective in giving rise to an ESR signal. This geometric requirement restricts the size of a light source which may be used. Generally, the spark gap or effective length of the discharge should not be greater than 3 cm. Proper utilization of focusing lenses and reflective mirrors can maximize the light input to the sample.

ESR spectrometers are justly notorious for being easily susceptible to rf interference generated by the electrical discharge in a flashlamp system. Precautionary measures should include suitable shielding of the flash apparatus (e.g., a Faraday cage) as well as rf filters on all power cords and interconnecting cables. In addition, all apparatus should be connected by heavy copper braid to a good ground. It is the authors' experience that pulsed lasers are especially troublesome in this regard. However, by placing the laser in the next room and using a Faraday cage and rf filters on all cables, we have been completely successful in removing all laser-generated interference.

In the authors' laboratory, two xenon flashlamps have been used with success (Novatron 185A and Suntron 6A, both manufactured by the Xenon Corporation). Both lamps dissipate a maximum of 100 J; however, the Novatron 185A has a flash duration of 100 μsec compared with the Suntron 6A's duration of 6 μsec. Both lamps have arc lengths of approximately 4 cm. The 185A lamp has a horizontal light output per

flash of ~100 mJrad^{-1}, with a maximum hold-off voltage of 2 kV. The
Suntron 6A is rated at 10 kV and has a light output of ~25 mJrad^{-1}.
Although the output from the 6A is less than that of the 185A, the
much shorter flash duration of the 6A ensures a higher maximum
light intensity. The 6A has an ingenious design feature which im-
proves reproducibility and increases the lamp's usable lifetime.
This technological advance consists of the incorporation of ballast
volumes around the anode and cathode. These ballasts act primarily
to absorb the plasma shock created during the lamp discharge. In
addition, they significantly reduce the rate of contamination.

     Housings suitable for accommodation of these flashtubes are
illustrated in Figs. 1 and 2. These lamp housings were constructed
from nonmagnetic materials, thus permitting the lamp to be placed a
minimal distance away from the sample. The flashlamp was positioned
between a quartz lens (2 in. focal length, 2 in. diam.) and a front-
surface aluminized spherical mirror (1 in. focal length, 2 in. diam.).
The mirror was adjusted so that its focal point coincided with the
center of the flashlamp. Hence the arc and its superimposed image
could then easily be focused on the sample cell by means of the

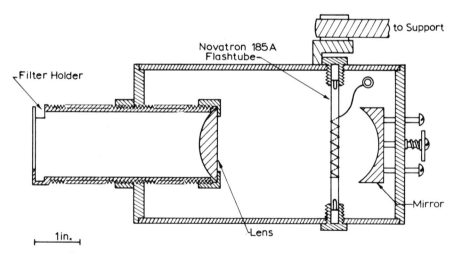

FIG. 1.   Flashlamp housing for the Novatron 185A (Xenon
Corporation).   After Hales and Bolton [17].

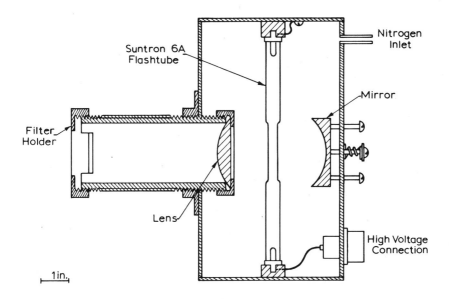

FIG. 2. Flashlamp housing for the Suntron 6A (Xenon Corporation). After Hales and Bolton [17].

front lens. Provisions for filters were also included on each focusing sleeve.

Use of an ellipsoidal reflector has also been reported [9].
The flashlamp is situated at one focus and the ESR sample tube at
the other. Lamp housings utilizing this geometry have considerably
greater light-gathering efficiency than conventional lamp geometries,
which is an important asset when lamps of short flash duration are
encountered. A recent innovation in housing and lamp design features
the addition of a fused silica lightpipe as the light carrier [10]
(see Fig. 3). A John Hadland HL 101 flash unit was modified to
permit transmission of the UV generated from the xenon plasma discharge. Tungsten electrodes with a high-purity xenon gas fill (100
Torr) provided a 3.5 μsec flash transient during discharge of 50 J
at 10 kV. A spherical mirror acted as a light collector and focused

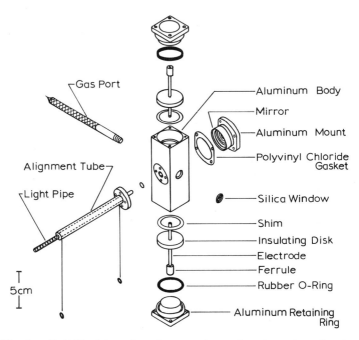

FIG. 3. Modified housing for a John Hadland HL 101 flashlamp.
The end of the light pipe is incident onto the ESR cavity. After
Doetschman [10].

the arc image onto the end of the lightpipe. Output from the light-
pipe-flashlamp system was 2.9×10^{16} photons.

The recent introduction of flashlamp-pumped dye lasers [11] to
the arsenal of photolytic sources heralds a new era in FPESR applications. Tunability and monochromatic output are the major attractive
features of dye lasers. However, dye laser conversion efficiencies
compete favorably with those of solid-state lasers, hence high output
powers (megawatts) are also available. Gross wavelength tuning can
be accomplished by simply changing dye solutions. In addition, fine
wavelength selectability over the dye emission range can be obtained
by the addition of a Fabry-Perot etalon or a grating as the dispersive
device. At present, flashlamp-pumped dye lasers are capable of yielding monochromatic radiation over a range of 350-700 nm [12]. Frequency
doubling by the addition of a non-linear crystal extends dye laser
capability into the UV.

A Synergetics Chromabeam 1070 flashlamp-pumped dye laser has
recently been interfaced with the ESR spectrometer (Varian E12) in
the authors' laboratory. Utilizing a combined coaxial flashlamp-dye
cell module, the Chromabeam 1070 is rated at 250 mJ output (Rhodamine
6G) with a maximum input of 100 J. The lamp rise time is approximately 200 nsec with a flash duration of 800 nsec. Lasing pulses
are typically 400 nsec in length and are repeatable at a rate of
20/min. Tunability is achieved by the addition of a Littrow
mounted grating, ruled at 1800 lines/mm and blazed at 5500 Å. This
grating makes possible spectral narrowing of the dye emission to 10 Å.

Pulsed nitrogen lasers have been applied to FPESR investigations
[13]. However, although their high repetition rates are advantageous,
the low output power can be very frustrating. The future application
of other laser sources to FPESR, particularly Nd or ion lasers in
tandem with Q-switch or Pockels cell control, will provide greater
flexibility to an already versatile technique.

B. ESR Spectrometer

Sensitivity, stability, and time response or effective bandwidth) are the three aspects of spectrometer design crucial to FPESR.
Modern solid-state ESR spectrometers approach the theoretical limit
of sensitivity. Any ESR spectrometer which does not achieve this
standard will not be very suitable for most FPESR studies as poor
signal-to-noise ratios are the rule rather than the exception.

Most modern spectrometers have field control systems which maintain the field to ±10 mG over a period of about 1 min. This is usually adequate when the ESR linewidth is greater than 1 G. However,
for narrower lines (as are often found for free radicals in solution),
long-term drifts in the field can cause an artificial broadening in
the CAT or even loss of the signal. The solution is to use a field-
frequency lock system such as the Varian E203 accessory. Long-term
drift is then reduced to about 3 mG. Most ESR spectrometers employ
phase-sensitive detection of the signal by modulating the magnetic

field at 100 kHz. Hence a typical bandwidth for such a spectrometer
is about 10 kHz or a response time of about 100 μsec. Thus a conventional spectrometer cannot detect free radicals with lifetimes
shorter than ~100 μsec.

Numerous changes must be instituted to reduce the response time
to about 1-2 μsec. One apparently simple solution is to increase
the modulation frequency; however, this entails certain problems.
As the modulation frequency is increased, the skin depth decreases
for penetration of the modulation field into the cavity. This problem can be circumvented by using a hairpin loop for modulation inside the cavity or by reducing the thickness of the cavity walls.
The first approach has been taken by Bennett et al. [9] with a modulation frequency of 465 kHz and a response time of 7 μsec. Smaller,
Avery, and Remko have also reported the use of a loop to introduce
2 MHz modulation with a response time of 3 μsec [14].

In contrast Atkins, McLauchlan, and Simpson utilized a thin-
wall Decca MW 232 cavity containing conventional Helmholtz coils [13].
These coils are driven at 2 MHz through an impedance-matching unit
(see Fig. 4). The maximum modulation field obtained at 2 MHz is 1 G
with a response time of about 1 μsec. Recently these investigators
have replaced the conventional cavity and impedance-matching unit
with a cavity of sufficiently thin walls that 2 MHz modulation coils
can be mounted on top of the standard 100 kHz coils. This arrangement provides approximately a 5 G (peak-to-peak) modulation field
from a 10 W input.

Although the rectangular TE_{102} cavity fitted with a slotted
irradiation grid is universally popular in both steady-state and
flash photolysis applications, other cavity configurations may prove
to be useful. The Varian E234 optical transmission cavity (TE_{103})
is particularly useful when negligible attenuation of the incident
light energy is desired. The optical transmission cavity consists
of two open windows located at the end of microwave cutoff extensions. It should be noted that the advantage gained in increased
light transmission by the E234 is offset by the consequent reduction in sensitivity which results from the unusual cavity geometry.

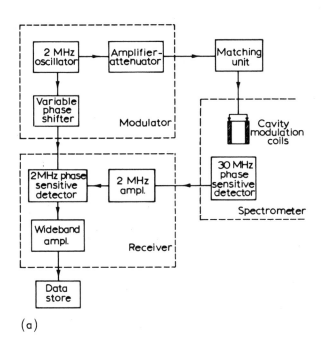

(a)

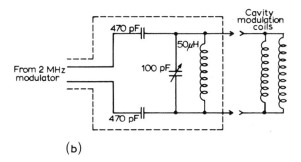

(b)

FIG. 4.   (a)   A block diagram for the 2 MHz detection system
reported by Atkins, McLauchlan, and Simpson. Details of operation
are given in Ref. [13].   (b)   Schematic for a 2 MHz impedance matching unit for the Helmholtz modulation coils. After Atkins et al.
[13].

The unloaded Q of the E234 is roughly two-thirds that of the standard TE_{102} cavity (Varian E4531). The E234 optical transmission cavity is superbly suited for double flash as well as simultaneous flash and steady-state illumination applications, since it provides for illumination from either direction.

C. Computer of Averaged Transients

Like conventional flash photolysis, FPESR requires a photolytic source, a flash power supply, and a trigger unit. For acquisition of data, however, transient ESR spectroscopy utilizes a computer of averaged transients (CAT). Unlike conventional flash photolysis, the transient signals observed in ESR are generally too noisy to be monitored by a single trace on an oscilloscope; hence repetitive techniques are a must. The CAT is basically a memory device containing 2^8-2^{12} memory channels. The signal to be acquired by the CAT is converted by the computer into a digital form which is then available for further manipulations. The primary advantage of the CAT is that the acquisition act can be repeated, with the digital results of the various scans successively summed. In this way, coherent signals will accumulate directly with the number of events n; however, incoherent noise will increase by only $n^{1/2}$. Hence the overall improvement in signal-to-noise is $n^{1/2}$.

Two basic modes of operation are common in FPESR instrumentation: (a) the kinetic mode and (b) the survey mode. The kinetic mode requires that the spectrometer magnetic field be fixed. The photolysis event is synchronized with the time scan of the CAT in order to provide an initial baseline to assist in kinetic analysis. The photolysis act is then repeated until a sufficient signal-to-noise ratio is obtained. The only information available from this procedure is the kinetic behavior of the transients of interest. However, by a "random" systematic variation by regular increments of the magnetic field, a composite of discrete kinetic traces can be obtained. The sum of these separate experiments can be manipulated by computer or manually to generate the spectra of the transients at any time after the flash. Hence, by a modification of the kinetic mode, maximal information about the photochemical reaction can be obtained, even though the task is somewhat laborious. In this mode, most of the information is in the time domain. Usually only a limited number of different field positions can be sampled; hence resolution of the reconstructed spectrum may be poor.

The survey mode primarily allows observation of the transient spectrum at a given time after the flash. In effect, the survey mode is a snapshot of the ESR spectrum at a given point in time. As such, this procedure generally does not yield kinetic information, but is principally useful in the identification of radicals present. The survey mode has been approached from two separate methodologies. The first procedure utilizes a CAT slaved to a slow field-scanning spectrometer. The second approach involves the rapid scanning of the magnetic field with the aid of Helmholtz coils affixed exterior to the microwave cavity. Both methods will be considered in greater detail below.

D. Rapid Scan

As indicated earlier, it is often desirable to have the complete ESR spectrum of radicals produced immediately after photolysis. This spectral profile can be obtained by a point-by-point method from kinetic studies by shifting the magnetic field for each kinetic run. This method gives only a low-resolution spectrum at best. Alternatively, one can use the technique of rapid scanning of the magnetic field to obtain a good spectral view. Although a number of reports in the literature refer to the use of such a rapid-scan technique [15-17], the results that have been obtained have not been spectacular and the details of experimental and design techniques are limited.

The successful utilization of the rapid scanning technique demands that the signal of interest be field scanned in a time period much shorter than the intrinsic lifetime of the radical. This requirement ensures that the spectrum obtained will not be distorted due to significant decay of the free radical during the rapid field sweep. In addition, it is essential that the field sweep be initiated at the same instant following each flash, thus preventing a distortion of the signal due to any possible nonreproducibility of the timing circuit. It is also desirable to have the field sweep be as linear as possible, since this facilitates calibration of the rapid scan and interpretation of the unknown spectrum. Homogeniety of the field sweep is also important because many organic radicals in solution often have linewidths of the order of 0.1 G.

Helmholtz coils are known to yield homogeneous fields and have been popular for rapid-scan applications. Design parameters for the coils can be simplified by consideration of the basic equation applicable to Helmholtz coils [18]:

$$H = \frac{B}{\mu} = \frac{8nI}{5^{3/2}a}$$

where H is the magnetic field (in gauss), B is the magnetic induction, and μ is the free-space permeability ($4\pi \times 10^{-7}$ H/M); n is the number of turns of wire per coil, and I is the current in amperes. The value of a in meters represents the radius of a coil and also the separation between the two coils.

The desired field range in gauss to be swept will directly determine the value of nI, the "ampere-turns," required. The value of nI will also be limited by the maximum current available from the amplifier driving the coils. The value for a should also be consistent with the "ampere-turn" number selected for the coil, and with the local geometry of the cavity and magnet pole faces. The coils must be of suitable size to allow facile installation between the pole face and the microwave cavity. However, it must be noted that homogeniety is enhanced by large values of a, hence both geometrical and homogeniety considerations are important.

The usual rapid-scan apparatus employs a sawtooth or ramp generator as the input to a power amplifier which drives the Helmholtz coils, mounted on both sides of the microwave cavity. A trigger

unit provides the synchronization necessary for starting the field
sweep, thus ensuring reproducible scans.

The initial report of a rapid-scan device by Sohma, Komatsu,
and Kanda in 1968 utilized oscilloscope recording and a single-shot
sawtooth signal, amplified and applied to Helmholtz coils [15]. The
coils were constructed of 150 turns of 0.6 mm diam copper wire (coil
o.d. of 40 mm and i.d. of 26 mm). Sweep range was approximately
250 G with a current input of 7 A. However, linearity was claimed
only over a range of 50 G. Few construction details were given.
The rapid-scan spectrum of the decaying semiquinone radical is illustrated in Fig. 5.

Similarly, Hirasawa et al. applied the rapid-scan technique to
a study of electrode processes by ESR [16]. These authors also utilized a sawtooth current as the driving waveform; however, the spectrometer output from the phase-sensitive detector was fed directly
into a CAT, thus allowing for signal averaging. Details of the instrumentation were omitted in this account except to indicate that
the coil consisted of 50 turns of 1 mm copper wire. Scan range was
variable from 0 to 100 G, and sweep time was independently variable
from 1 to 50 msec.

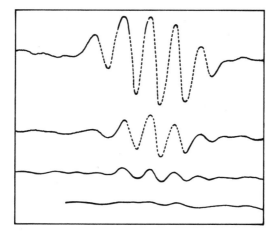

FIG. 5. Rapid-scan presentation of semiquinone during decay.
Each spectrum was acquired in 20 msec. The interval between spectral
traces was 1.5 sec.

The initial impetus to construct a rapid-scan system in the authors' laboratory resulted from the need to have adequate concentration data to allow calculation of second-order rate constants for biologically significant semiquinone disproportionations. The preliminary apparatus employed a unijunction transistor as the sawtooth generator and a Kepco operational amplifier (OPS21-1B) as the power amplifier to drive the Helmholtz coils [17] (see Fig. 6). The frequency of the sawtooth waveform was variable from 1 to 400 Hz, thus allowing for a multiple field sweep during the single time sweep of the CAT. The field sweep range was variable from 1 to 40 G. The sawtooth was synchronized with the triggering unit providing reproducible firing of the photolysis flash after the initiation of the sawtooth. Likewise, the CAT time scan and the sawtooth waveform were coincident. Calibration of the sweep was achieved by using the p-benzoquinone anion (10^{-4} M).

Although the prototype system performed admirably for the most part (Fig. 7), certain limitations have prompted major alterations

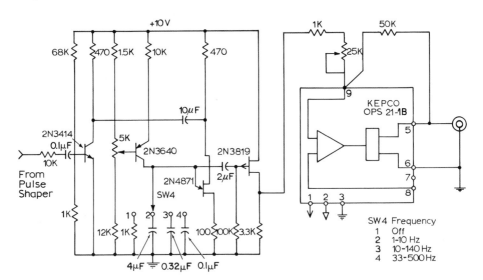

FIG. 6. Output stage of a rapid scan apparatus reported by Hales and Bolton [17]. The initializing spike from the pulse shape triggers the sawtooth forming circuit. The output from the Kepco OPS21-1B is fed directly into the rapid scan coils.

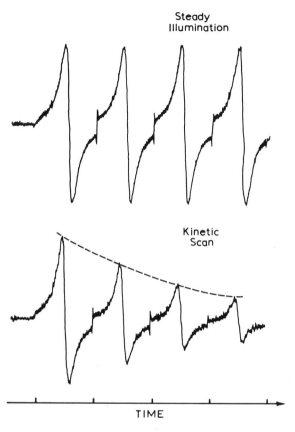

FIG. 7. Somparison of the rapid-scan spectrum of the chloranil
anion during steady-state illumination and after flash photolysis.
After Hales and Bolton [17]. Each unit on the time base is 20 msec.

in the basic apparatus. The early system possessed a pronounced
nonlinearity during the initial 15% of the field sweep; therefore,
it was necessary to position the resonance in the linear section of
the rapid field sweep range to prevent distortion of the acquired
signal. Likewise, the field sweep range was of limited width, and
the prototype coils were small, leading to field homogeniety problems.

It has been well documented that the voltage waveform that will
best produce a linear current ramp in an RL circuit is trapezoidal

[19]. Therefore, instead of the standard sawtooth ramp as the field
sweep voltage, a trapezoidal voltage form has been recently employed
[20]. This waveform has been obtained by summing two waveforms generated by a Fabri-tek 1072 CAT: a linear ramp (0-4 V) and a square wave
(0-3 V). The resulting current ramp is made linear by varying the
amplitude of the square wave added to the CAT voltage ramp (see Fig.
8). Linearity better than 4% has been routinely achieved. The
linear voltage ramp generated by the CAT has another important use.
Since this ramp voltage is directly proportional to the channel
number in the CAT memory, the rapid scan sweep (i.e., the current
ramp) is therefore synchronized with the time sweep of the CAT.
Hence an initial trigger pulse leads to the simultaneous scan of
the CAT through its memory channels and a linear sweep of the field.

Since the current limit of our driving amplifier (Kepco OPS21-1B)
is 1 A, severe restrictions have been placed on our coil design. As
a compromise, our sweep width is arbitrarily chosen to be 40 G. To
satisfy the geometrical considerations peculiar to our spectrometer
(12 in. magnet) and to ensure that the coils will be easily mountable,
the coil radius has been fixed at 4 cm. These considerations applied
to the basic Helmholtz coil equation yield an "ampere-turn" number
per coil of about 200. The coil mounts, which are made of Plexiglas,

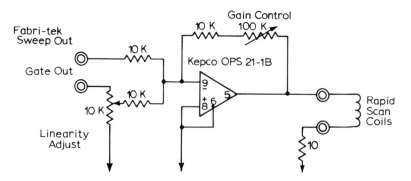

FIG. 8. Output stage for a rapid-scan apparatus utilizing waveforms generated by a Fabri-tek 1072. Sweep width (in gauss) is variable and is selected by the gain control. After Hsi, Fabes, and
Bolton [20].

bolt onto the magnet pole faces. Each coil consists of 200 turns
of #26 AWG (American Wire Gauge) enamel coated wire. Routinely we
are able to sweep 40 G in 5 msec with a spectral resolution better
than 0.1 G. Faster sweep times are not possible with our present
apparatus, because of limitations imposed by the dwell time of our
CAT (20 μsec per channel) and the time constant of our spectrometer.

Although rapid scanning techniques have been applied primarily
to photochemical systems, their utilization in the study of other
transient events by ESR (i.e., pulse radiolysis, pulse electrolysis)
should prove to be beneficial in unraveling the mechanisms of electron transfer and electrode processes.

E. Flash Photolysis-Spectrometer Interface

The heart of the typical FPESR apparatus is unquestionably the
trigger or sequencer unit. This component provides synchronization
for starting the CAT scan, triggering the photolyzing source, and,
in some applications, starting the rapid scan or advancing the magnetic field setting by a small increment.

The trigger unit (Fig. 9) in use at our laboratory (the "pulser")
is a versatile instrument suitable for a number of flash sources (including our dye laser), rapid-scan applications, and external switching of a photomultiplier. The pulser initiates lamp (or laser) discharge through an SCR switch which is normally maintained at 200-300
V (Fig. 10). A small voltage trigger pulse (2 V) generated by the
pulser internal timer dumps the SCR to ground potential, allowing a
brief voltage spike (1 μsec) to be delivered to a standard auto ignition coil. The resulting 20 kV output of the ignition coil is
sufficient to trigger discharge of the xenon lamps of the laser spark-
gap. The time between trigger pulses is continuously variable from
1 to 45 sec, thus allowing for a wide range of repetition rates.

To provide an initial baseline for kinetic studies, the trigger
impulse to start the CAT scan is generated before the lamp discharge

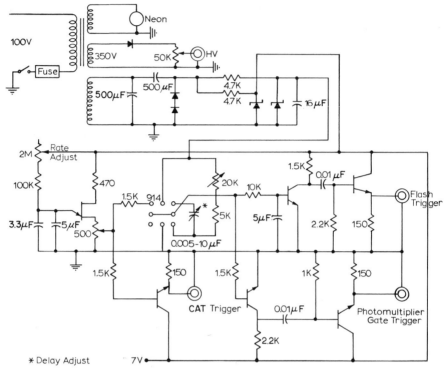

FIG. 9. Schematic for the sequencer unit utilized in the authors' laboratory. BNC outputs provide sequencing for CAT sweep initialization, triggering for the photolytic flash, and external switching of a photomultiplier.

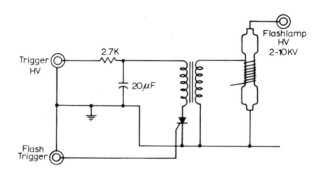

FIG. 10. Schematic representation of lamp triggering circuit. Trigger HV and flash trigger pulse are generated by the sequencing unit. A standard auto ignition coil provides the 20-50 kV spark needed to initiate lamp breakdown.

pulse is applied. The delay between the CAT trigger and the lamp
trigger is variable from 10 μsec to 250 msec, thus accommodating a
variety of CAT acquisition times.

The accompanying block diagram (Fig. 11) serves to illustrate
the interrelationship of the basic components in our FPESR apparatus.
The pulser unit provides the necessary synchronization between the
CAT, flashlamp, and spectrometer. The resulting kinetic data or field
profile stored in the Fabri-tek 1072 is available for oscilliscope display or readout on an x-y recorder. Alternatively, the CAT memory
contents can be processed by a Fabri-tek 281 paper punch control and
then relayed to a Tally tape perforater for storage on paper tape and
subsequent computer analysis.

Other approaches to data collection from FPESR spectrometers
have been utilized with varying degrees of success. Firth and Ingram
provided synchronization of the spectrometer field sweep with the repetitive sequential sweep of the CAT through its memory channels (Fig.
12) [21]. Hence each channel represented a small increment of field.
The stimulus to produce free radicals was applied at a repetition
rate greater than the field sweep repetition rate. A variable delay
was imposed after each stimulus before the CAT was allowed to sample

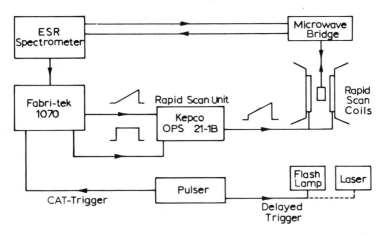

FIG. 11. Block diagram for a Flash Photolysis ESR spectrometer.
Greater detail of the rapid scan attachment is given in Fig. 8.

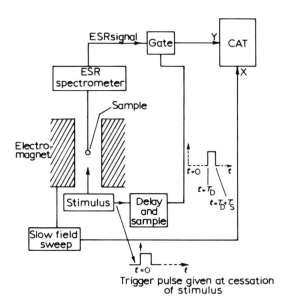

FIG. 12. Diagramatic representation of the transient ESR spectrometer designed by Firth and Ingram [21].

the incoming ESR signal. Thus the field profile of the radical could
be viewed at any desired time after the stimulus. Additionally, the
CAT imput was gated to prevent acquisition of spurious signals and to
achieve short sampling times. It should be noted that the excitation
source was not synchronized with the field sweep; hence the proper
choice of stimulative repetition rate was essential in ensuring that
all CAT channels were sampled equally.

    An alternative method, reported by Bennett, Smith, and Wilmshurst
employed a slow field sweep with six sampling channels available for
data acquisition (Fig. 13) [9]. Suitable delays were imposed, allowing each channel to sample a particular time after the photolyzing
flash. Thus spectra were available from six points on the decay curve.
Each channel possessed a 1.5 μsec sampling time, and the sampling sequence was synchronized to the flash by a photodiode. The resulting
1.5 μsec signals from the spectrometer were stretched by two cascaded
diode capicitors, and after suitable gating, were fed into a pen recorder for display.

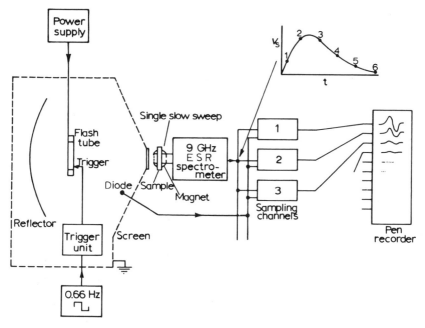

FIG. 13.   The flash photolysis ESR interface reported by
Bennett, Smith, and Wilmshurst [9].

Successful interfacing of a pulse-radiolysis source to a homo-
dyne ESR spectrometer has been demonstrated by Smaller, Avery, and
Remko [14]. Detection of transient species having lifetimes on the
order of a few microseconds is achieved by acquiring the kinetic
data with a 100 channel analog memory device. Processing of the raw
data is permitted by transfer of the analog memory contents to a
digital memory unit. A field profile of the decaying species is obtained by substituting a boxcar integrator for the analog device.
The digital memory produces the necessary incrementation of the field
to yield the final spectrum.

With a Biomac 500 data storage instrument having an auxillary
sweep expander to provide channel dwell time of 0.1 μsec, the 2 MHz
detection system of Atkins, McLauchlan, and Simpson allows transient
radicals with lifetimes of a few microseconds to be observed [13].
Three modes of spectrometer operation have been detailed by Atkins

et al. Mode I primarily provides the free-radical decay curve obtained at a fixed magnetic field. In contrast, Modes II and III
generate the spectrum of the decaying species at a particular time
after the actinic flash.

Mode I [Fig. 14(a)] corresponds to the kinetic mode discussed
earlier. The high repetition rate of the N_2 laser provides a usable
signal-to-noise ratio for samples of low concentration or signal
amplitude. Mode II operation [Fig. 14(b)] is achieved by synchroni-
zing the field sweep with the scan and sampling unit. After all
addresses have been sampled, the field and CAT sweep can be repeated
until a proper signal-to-noise ratio is obtained. Typically the laser
is operated at approximately 20 pulses/sec; yielding a total field
sweep time of 20 sec (for a 400 channel CAT). For suitable resolution
of the hyperfine splitting, the field must be quite stable and the
successive sweeps reproducible. This method is slow and is best
adapted to signals of strong intensity. Mode III [Fig. 14(c)] com-
pensates for the disadvantages associated with spectrum acquisition
by mode II. The field is now swept very slowly (1 scan per 100 min);
however, the laser is pulsed at its maximum rate (100 Hz). The sample and delay unit is used as in mode II; however, the address advance of the CAT is inhibited until the signal-to-noise ratio is at
a usable value. The acquisition time for each channel is preset and
depends on the strength of the transient signal. Mode III yields a
better signal-to-noise ratio than mode II, as a consequence of the
higher sampling number per address. In addition, acquisition time
is reduced in mode III; however, it should be noted that this method
is practical only with light sources capable of high repetition rates.

III. APPLICATIONS

Although FPESR has an excellent potential for the elucidation
of photochemical mechanisms, the recent literature is virtually de-

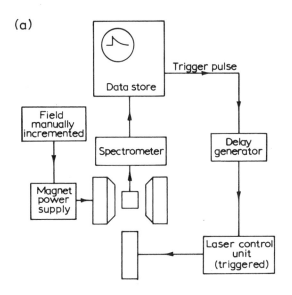

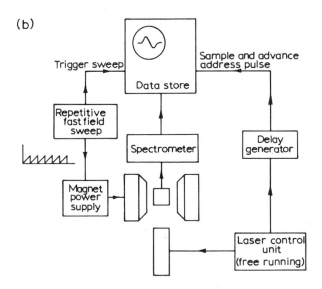

FIG. 14.   Block diagram of the operating modes of the FPESR apparatus constructed by Atkins, McLauchlan, and Simpson [13]:   (a) mode I, (b) mode II, (c) mode III.

(c)

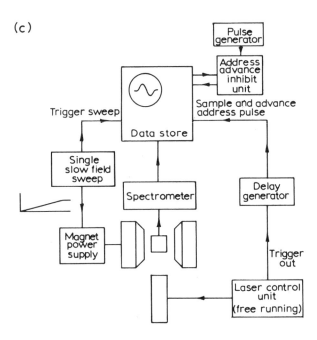

void of FPESR investigations. This can perhaps be explained by the relative novelty of the technique. The few communications published mainly concentrate on chemical reaction systems; however, certain biological systems are also suitable for study by transient ESR.

The majority of the exploratory investigations using FPESR have focused on the behavior of the benzophenone radical anion formed by UV photolysis. The pioneering studies of Bennett, Smith, and Wilmshurst [6] provided confirmation for the earlier optical data of Beckett and Porter [22]. Although Bennett et al. had sufficient instrumental time response (7 μsec), the low signal intensity after photolysis did not permit resolution of the hyperfine pattern (Fig. 15). Subsequent refinements of the signal acquisition technique were reported by Atkins, McLauchlan, and Simpson [23] and a "high-resolution" spectrum of the benzophenone radical anion was exhibited. Later investigations of the same chemical system using laser photo-

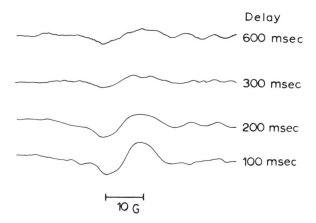

FIG. 15.   Spectrum of the benzophenone radical anion after photolysis as a function of the delay time after the flash.   After Bennett, Smith, and Wilmshurst [6].

lysis and a 2 MHz detection system provided evidence that during the first 100 μsec of lifetime, the neutral ketyl radical $(Ph_2COH\cdot)$ ESR spectrum was exhibited in emission [13] (Fig. 16).   The relaxation time (T_1) for this radical in solution was determined to be approximately 40 μsec.

The photochemical behavior of chlorophyll and quinone in solution has been of considerable interest in our laboratory, since both of these molecules are involved in the electron transport reactions of photosynthesis.   Our interest in this system served as an impetus to build a flash photolysis-ESR interface, because it was known from the work of Tollin et al. that semiquinone free radicals are intermediates in the photochemical reactions [24].

Before work was started on the chlorophyll-quinone system, a study was made of the photolysis of chloranil as a test system [17] since the photochemistry and kinetics were well established [25]. From this system, we developed techniques for spin concentration measurements of transients and rapid-scan techniques.

The chlorophyll-quinone system as studied by FPESR proved to be much more complex than had been surmised from optical flash photolysis work and steady-state ESR.   The kinetics of the decay of the

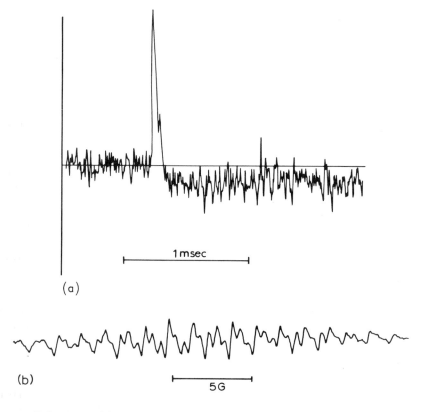

FIG. 16. (a) Kinetic trace showing the behavior of the benzophenone radical anion during the first millisecond after photolysis.
The initial sharp spike represents emission from the excited radical.
(b) Spectrum of the neutral ketyl radical ($Ph_2COH\cdot$). After Atkins,
McLauchlan, and Simpson [13].

p-benzosemiquinone anion were second order. The rate and activation
energy as determined by FPESR agreed with earlier work by Tollin et
al. using a shutter technique [24]. However, when the modulation
amplitude and microwave power were significantly increased, a new
transient appeared with a half-life of about 0.5 msec (Fig. 17).
This new transient decayed via first-order kinetics. When perdeutero-
quinone was employed, the linewidth narrowed considerably, proving
that the transient contained the semiquinone. Since the free

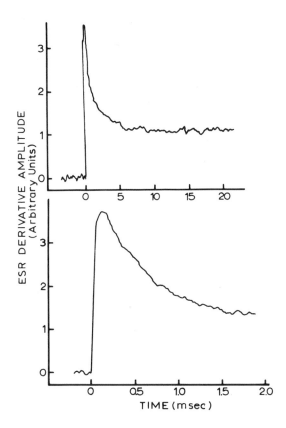

FIG. 17. Decay of a new transient intermediate in the chlorophyll a-p-benzoquinone system. The upper trace illustrates the super-
position of the rapid decay of the new species on the slower decay of
the p-benzoquinone anion. The lower trace is an expansion of the
initial portion of the upper trace. After Hales and Bolton [17].

semiquinone is also seen separately, this new transient is tentatively
identified as a chlorophyll cation-semiquinone biradical complex [26].

Application of the FPESR method to assist in the decoding of
significant biological problems has only been recently reported.
Utilizing straight detection from the microwave detector crystal,
Dutton, Leigh, and Seibert were able to monitor the fast formation
(∼40 μsec) of radical and triplet bacteriochlorophyll in subchromato-
phore particles prepared from the photosynthetic bacterium Chromatium

D [27]. The triplet spectrum is observed only when light-induced
electron transfer from the specialized bacteriochlorophyll P883 is
blocked. A portion of the triplet spectrum is in emission, signify-
ing that some of the triplet spin levels are selectively populated
and that the spin-lattice relaxation time is extremely long at the
cryogenic temperatures employed. It is worthy of note that straight
detection from the microwave crystal is possible only when excellent
signal-to-noise ratios are available, and since the dye laser photolysis of Chromatium D was carried out at liquid helium temperatures,
this condition was easily achievable.

As part of our research program in photosynthesis, we have been
probing the nature of light-generated electron flow in photosynthetic
bacteria, algae, and green-plant chloroplasts [28]. Although FPESR
has proved advantageous in this regard, correlation of transient
radical species with light-induced optical absorbency changes is
perilous because different samples under widely varying conditions
are often used. To avoid this pitfall, we have recently developed
an apparatus for simultaneously monitoring optical and ESR transients
after an actinic flash from a Synergetics Chromabeam 1070 dye laser
[29] (Fig. 18). Our Varian E12 has been fitted with an optical trans-

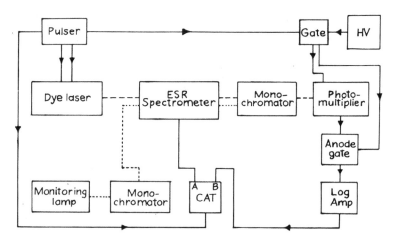

FIG. 18.   Block diagram for the simultaneous optical and ESR
transient spectrometer.

mission of the monitoring and photolytic light beams through the
cavity. Protection of the photomultiplier (EMI 9558A) from the actinic flash is achieved by the insertion of a Jarrel-Ash 0.25 m
grating monochromator immediately before the photomultiplier. In
addition, the photomultiplier high-voltage supply as well as the
photomultiplier output can be gated to coincide with the photolytic
excitation. The transient optical and ESR kinetics are simultaneously
acquired by our CAT.

The kinetic and quantitative correlation between a light-induced
ESR resonance and an optical bleaching at 700 nm in spinach subchloro-
plast particles has recently been demonstrated (Fig. 19) using the
simultaneous optical-ESR transient spectrometer [30]. This correlation strongly supports the contention that the narrow light-induced
ESR resonance represents the primary photochemical step in plant
photosynthesis.

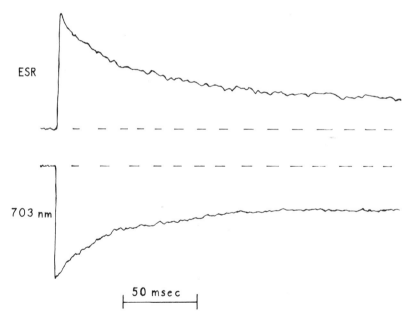

FIG. 19. Simultaneous optical and ESR decay kinetics of spinach D144 particles incubated with PMS (1 μM). Actinic wavelength 620
nm. The traces are an average of 64 sequential flashes. After
Warden and Bolton [30].

IV. CONCLUSION

Although flash photolysis electron spin resonance is still a
relatively untried and an emergent technique, we are confident that
this tool will play a major role in the future elucidation of fundamental photochemical processes. Thus advances in experimental
technique as well as instrumental design are to be expected as FPESR
is adopted by a wider range of investigators. It is our hope that
this introduction to the methodology of FPESR will convince the
practicing photochemist of the value of ESR in photochemical research, and in addition stimulate the ESR practitioner to investigate the possible applications of transient ESR spectroscopy in his
research.

ACKNOWLEDGMENTS

We would like to express our thanks to Dr. L. G. Fabes, Dr. B.
Smaller, Dr. E. A. McLauchlan, and E. S. P. Hsi for helpful discussions. This work was supported in part by a National Research Coun-
cil of Canada Bursary (to JTW). This chapter is Publication No. 39
of the Photochemistry Unit, Department of Chemistry, University of
Western Ontario, London, Canada.

REFERENCES

1. G. Porter, in Photochemistry and Reaction Kinetics (P. G.
    Ashmore, F. S. Dainton, and T. M. Sugden, eds.), Cambridge Univ.
    Press, London, 1967, pp. 93-111.

2. J. E. Wertz and J. R. Bolton, Electron Spin Resonance--Elementary
    Theory and Practical Applications, McGraw-Hill, New York, 1972.

3. H. M. Swartz, J. R. Bolton, and D. C. Borg, (eds.) Biological
    Applications of Electron Spin Resonance, Wiley, New York, 1972.

4. C. P. Poole, Jr., Electron Spin Resonance--A Comprehensive
    Treatise on Experimental Techniqes, Wiley (Interscience), New
    York, 1967.

5. R. S. Alger, Electron Paramagnetic Resonance: Techniques and
    Applications, Wiley, New York, 1968.

6. T. J. Bennett, R. C. Smith, and T. H. Wilmshurst, Chem. Commun.,
    307, 513 (1967).

7. M. L. Randolph, in Biological Applications of Electron Spin Resonance (H. M. Swartz, J. R. Bolton, and D. C. Borg, eds.), Wiley, New York, 1972.

8. J. R. Bolton, H. M. Swartz, and D. C. Borg, in Biological Applications of Electron Spin Resonances (H. M. Swartz, J. R. Bolton, and D. C. Borg, eds.), Wiley, New York, 1972, Chap. 2.

9. T. J. Bennett, R. C. Smith, and T. H. Wilmshurst, J. Phys. (E), 2, 393 (1969).

10. D. C. Doetschman, Rev. Sci. Instr., 43, 143 (1972).

11. B. B. Snavely, Proc. IEEE, 57, 1374 (1969).

12. J. T. Warden and L. Gough, Appl. Phys. Letters, 19, 345 (1971).

13. P. W. Atkins, K. A. McLauchlan, and A. F. Simpson, J. Phys. (E), 3, 547 (1970).

14. B. Smaller, E. C. Avery, and J. R. Remko, J. Chem. Phys., 55, 2414 (1971).

15. J. Sohma, T. Komatsu, and Y. Kanda, Japan. J. Appl. Phys., 7, 298 (1968).

16. R. Hirasawa, T. Mukaibo, H. Hasegawa, Y. Kanda, and T. Maruyama, Rev. Sci. Instr., 39, 935 (1968).

17. B. J. Hales and J. R. Bolton, Photochem. Photobiol., 12, 239 (1970); B. J. Hales, Ph.D. Thesis, Univ. Minnesota, 1970.

18. See Ref. 3, page 418.

19. S. Seely, Electronics Circuits, Holt, Rinehart, and Winston, New York, 1967, p. 626.

20. E. S. P. Hsi, L. Fabes, and J. R. Bolton, Rev. Sci. Instru., 44, 197 (1973).

21. E. W. Firth and D. J. E. Ingram, J. Sci. Instr., 44, 821 (1967).

22. A. Beckett and G. Porter, Trans. Faraday Soc., 59, 2038 (1963).

23. P. W. Atkins, E. A. McLauchlan, and A. F. Simpson, Nature, 219, 928 (1968).

24. D. C. Mukherjee, D. H. Cho, and G. Tollin, Photochem. Photobiol., 9, 273 (1969).

25. D. R. Kemp and G. Porter, Chem. Commun., 1029 (1969).

26. B. J. Hales and J. R. Bolton, J. Am. Chem. Soc., 94 (in press).

27. P. L. Dutton, J. S. Leigh and M. Seibert, Biochem. Biophys. Res. Commun., 46, 406 (1972).

28. J. R. Bolton, B. J. Hales, E. S. P. Hsi, and J. T. Warden, Abstr. Intern. Conf. Photosynthetic Unit, Argonne, Ill., 1970.

29. J. T. Warden, Ph.D. Thesis, Univ. Minnesota, 1972.

30. J. T. Warden and J. R. Bolton, J. Am. Chem. Soc., 94, 4351 (1972).

31. See J. G. Calvert and J. N. Pitts, Jr., Photochemistry, Wiley, New York, 1966, pp. 710-718, for a description of conventional xenon flash lamps.

CHAPTER 3

PICOSECOND LASER TECHNIQUES

Michael M. Malley

Department of Chemistry
San Diego State University
San Diego, California

I. INTRODUCTION

This chapter describes the experimental techniques for using mode-locked neodymium: glass and ruby lasers to study emission and absorption spectra on a picosecond time scale. "Mode-locked" in this context means that a periodic train of picosecond (10^{-12} sec) light pulses is emitted by the laser. Since the discovery of mode-locked laser oscillation in 1966, there have been a plethora of theoretical and experimental papers which discuss various aspects of picosecond laser pulses. The reader is referred to several review articles [1]. The emphasis in this chapter is on the experimental apparatus and operating procedures; no attempt is made to comprehensively review all applications which appear in the literature.

One might ask how the laser generates a train of picosecond light pulses. At the present writing, there is no theoretical model of passive mode-locking with a saturable dye (the most widely used method) that is able to satisfactorily explain the many experimental observations. Nevertheless, we can gain some qualitative understanding by considering the fundamentals of a laser. Briefly, these lasers consist of a solid rod of active medium, either neodymium-doped glass or ruby crystal, with the ends polished optically flat and parallel, and a Q-switch cell which contains a saturable dye. These elements are located between two dielectric-coated mirrors which are very accurately aligned as shown schematically in Fig. 1. These elements form the laser cavity which has an overall optical length L. The laser rod is surrounded by a helical flashlamp (or alternatively, it may be parallel to a linear flashlamp in an elliptical cavity) which is connected to a large high-voltage capacitor. When the flashlamp discharges, the light is absorbed by the rod and a population inversion of energy levels results. That is, there are more centers in the excited state than in the ground state, as shown in Fig. 2. There is spontaneous fluorescence from these excited states, and the emitted photons stimulate the other excited states to emit

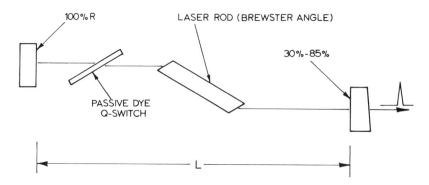

FIG. 1. Elements of a passive dye mode-locked laser. The laser rod is cut at the Brewster angle to eliminate extraneous reflections in the laser cavity. The overall optical length of the cavity is L.

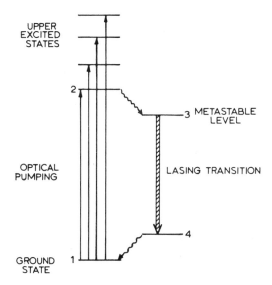

FIG. 2. Energy-level diagram of a four-level laser. The flashlamp excites the active centers to the upper levels, which then rapidly
relax to the metastable level. Lasing takes place between the metastable energy level and a lower level. In the case of ruby, the lower
level of the lasing transition and the ground state are identical,
i.e., ruby is a three-level laser. Neodymium:glass is a four-level
laser system.

photons with the same phase and direction. Naturally, there is also

absorption from the ground state by unexcited centers, but since

there are more centers in the excited state than the ground state

there is a net increase or amplification of light. This is of course

the meaning of the acronym LASER, namely, light amplification by the

stimulated emission of radiation. The flashlamp-pumped laser rod is

an amplifier for light in a certain band of wavelengths as shown in

Fig. 3. Making an analogy with electronics, one can consider that

the rod is an amplifier having a certain bandwidth and that the ex-

ternal mirrors are the feedback elements. When the gain of the amp-

lifier exceeds the losses of the laser cavity, then oscillation takes

place.

Within the bandwidth of the optical gain of the laser rod, there

are certain discrete wavelengths λ_i which fulfill the condition

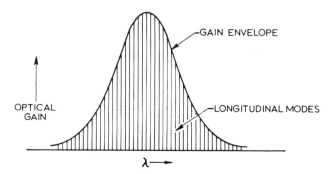

FIG. 3. Distribution of the optical gain of the active medium
over a band of wavelengths. Within the bandwidth are many discrete
longitudinal modes. The laser will oscillate in a particular mode
when the gain exceeds the losses at that wavelength. Mode-locked
laser output results when all the modes oscillate in phase.

$N\lambda_i = L$, where N is an integer and L is the optical length of the
laser cavity. Each of these wavelengths is a longitudinal "mode,"
that is, each of these represents a wavelength which suffers less
attenuation or loss than the wavelengths which do not fulfill the
above condition. In a sense, the external mirrors selectively feed
back to the amplifier these discrete wavelengths. When the gain in
the laser rod exceeds the losses for a given longitudinal mode, then
that mode will oscillate, and light will be emitted at that wavelength.

Lasers may also oscillate in different transverse modes. These
(discrete) transverse modes determine the light intensity across the
laser beam. The ideal would be a plane wave of uniform light intensity emitted by the laser. The closest approach to this ideal is
the lowest-order transverse mode TEM_{00}. For this mode, the light
intensity follows a Gaussian distribution as a function of the radial
distance from the center of the laser beam. The higher-order modes
consist of symmetrical patterns of light intensity with one or more
vertical or horizontal nodal planes.

For a typical passive-dye Q-switched mode-locked neodymium:glass
laser, L = 1 m, λ = 1.06 μ with an oscillating bandwidth of about 10
nm. This means that there are approximately 10^4 oscillating longitudinal modes. If we were to cause all of the modes to oscillate with

the same phase, then the resulting output is a periodic train of
ultrashort, bandwidth-limited pulses (τ = 100 cm^{-1} = 0.3 psec) separated in time by 2L/c, namely, the round-trip time in the laser cavity. The power of these ultrashort pulses is 10^4 times as great as
the total average power in all the longitudinal modes not oscillating
in phase. Thus the short-duration, high-power pulses are the result
of the constructive interference of all of the oscillating longitudinal modes of the laser.

The element which causes the modes to oscillate in phase is the
saturable-dye Q-switch. The properties of a dye which make it desirable as a mode-locking element are (a) a large extinction coefficient at the laser wavelength, (b) a fast relaxation time from the
excited state to the ground state (it is mandatory that the relaxation time be shorter than the cavity round-trip time), and (c) no
excited-state absorption of the laser light. These properties imply
that the absorbance will decrease, that is, dye will be optically
"bleached," at the light intensities appropriate for mode-locking,
typically in the range of 50 MW/cm^2.

One may view the evolution of the laser pulse train in the following way. Initially there is only spontaneous fluorescence or light
"noise" at the laser wavelength. As the gain of the laser rod increases and the light circulates in the cavity, it reaches a power
level such that the most intense fluctuation begins to "bleach" the
saturable-dye Q-switch. This means that this fluctuation suffers
less attenuation than the less intense light "noise" fluctuations
and therefore grows more rapidly than the less intense fluctuations.
The dye tends to shorten the pulse length and broaden the spectral
width of the pulse. When the pulse reaches power levels greater
than 500 MW/cm^2, the dye is completely bleached and the subsequent
pulse continues to spectrally broaden and also lengthens in time
duration because of dispersion in the laser rod. Finally, the excited-state population of the active medium is depleted, the gain decreases, and the pulse decays. Typically, 20-50 pulses will be observed in a pulse train, with a smoothly rising and decreasing
envelope.

There is general agreement that the duration and bandwidth of each of the subsequent pulses in the train of a mode-locked neodymium:glass laser increases. Initially the pulses have a bandwidth of 0.3 nm, and grow to a bandwidth of 8 nm at maximum intensity [2]. The typical values for the duration of pulses from a mode-locked neodymium:glass laser range from 3 to 12 psec, and powers from 1 to 10 GW. Within each of these picopulses there is a frequency sweep or "chirp." This has been measured by Treacy [3] and more recently by Auston [4]. The leading edge of the picopulse begins at longer wavelengths, and there is a gradual sweep toward shorter wavelengths as the pulse progresses.

It would certainly be desirable to obtain mode-locked behavior with bandwidth-limited picopulses in the neodymium:glass laser. If the entire bandwidth of spontaneous fluorescence ($\sim$240 cm^{-1}) could be perfectly mode-locked, then pulses of duration equal to 0.06 psec would be generated. With the advent of dye lasers, especially the recently reported continuous mode-locked dye lasers [5,6], it would appear that soon short-duration pulses will be available at any wavelength in the visible and near UV. Arthurs, Bradley, and Roddie [7] have reported transform-limited frequency tunable pulses in a flashlamp pumped rhodamine 6G dye laser.

The ultimate short pulse made up of visible wavelengths will be of the order of 1 femtosecond (10^{-15} sec), corresponding to one-half cycle of light. But since this pulse contains all visible wavelengths, one will not be able to determine to which state the molecule has been excited.

There is a qualitative difference between ordinary "fast" instrumentation [8] for studying excited-state lifetime and time-resolved picosecond spectroscopy. Presently, commercially available electronics are limited to risetimes of 0.3 nsec or longer. This is some two orders of magnitude slower than the light pulses from a mode-locked laser. One must invent new techniques for this time scale and, in particular, one must exploit the nonlinear interactions of light which occur with these ultrashort high-power picopulses. It is particularly apt in this time scale to translate time into distance,

using the speed of light as the conversion factor. Consider that
in air light travels 0.3 mm/psec, or 0.2 mm/psec in a medium with
an index of refraction of 1.5. One can see that this makes a very
convenient scale for studying ultrafast kinetics.

The first direct measurement of the duration of picopulses uses
the nonlinear interaction of light with molecules and translates time
into a spatial scale in a very ingenious manner. This is the now
classic two-photon excitation of fluorescence technique first demonstrated by Giordmaine, Rentzepis, Shapiro, and Wecht [9]. This experiment is described in more detail in the following discussion of
pulse measurements.

All of the experiments involving time-resolved picosecond spectroscopy utilize these basic concepts directly or indirectly. Both
time-resolved emission and absorption experiments have been performed
with picosecond pulses. In the following sections we describe techniques for setting up a Q-switched mode-locked laser with the associated accessories for the detection of mode-locked behavior, the
nonlinear techniques to generate picosecond pulses at different wavelengths, and finally various experiments which have been performed
to measure picosecond kinetics.

II. LASER SYSTEMS

We discuss in some detail how to assemble and operate a mode-
locked laser system and the associated optical accessories. Assembling a system from subunits has the advantage of a considerable
savings in cost and an inherent flexibility which cannot be achieved
in a complete commercial unit. A good source of information is the
buyer's guide published annually in Laser Focus. This is a comprehensive list of suppliers of components and manufacturers of complete
laser systems.

Basic to any system is an optical table. Many variations are
possible, commercial and otherwise. The requirements are that the

table be approximately flat and rigid. Vibration isolation is not
necessary unless one plans to do experiments in holography. There
must be some means of fastening the various laser and optical components firmly to the table. One technique is to use a large steel
plate ¼-½ in. thick, ranging in area up to 4 x 10 ft, supported
on a table. The laser and associated optics are then mounted on
magnetic bases which are available from machine-tool companies or
Edmund Scientific at very reasonable cost. This offers complete
flexibility in the placement of the various components, and the
bases are easily moved by "turning off" the magnetic hold-down. The
other common technique is to use a table with holes drilled and tapped
on a grid network covering the surface. The bases for components are
then held in place by using "dogs," i.e., step clamps, which are fas-
tened to the table with bolts.

The laser power supply and flashlamp head can be purchased from
any of several laser companies. The basic model K-1 sold by Korad
(a division of Hadron, Inc.) is a very popular laser unit. Since
there is strong competition among various laser companies, the cost
of components is fairly reasonable. A helical flashlamp surrounding
the rod is preferable to a linear flashlamp because the optical pumping is more uniform. If linear flashlamps are used, a diffuser of
frosted glass should surround the laser rod to provide a more uniform
flux of light and avoid "hot" spots. Water cooling of the laser rod
and flashlamp provides a more rapid return to thermal equilibrium
than air cooling. It is still necessary to wait approximately two
minutes after each laser shot in order to obtain the best results.

Owens-Illinois and American Optical both make neodymium:glass
laser rods to specifications. Union Carbide, Crystal Products Division, and Crystal Optics Research fabricate and finish ruby laser
rods to order. The recommended doping for neodymium:glass rods is
3% by weight. There is no general consensus about the doping for
ruby crystals, both 0.03% and 0.05% doping have been used for mode-
locked experiments. The higher doping means higher gain which is
important for the small-aperture single-transverse mode (TEM_{00}) laser
output.

A laser rod with both ends cut at the Brewster angle is preferred for several reasons: it eliminates extraneous reflections in the laser cavity (which can destroy mode-locking); there is no loss for light traveling into and out of the rod; and in the case of neodymium:glass, it provides polarized light output (normally there is no preferred polarization in neodymium:glass). Careful inspection of the laser rod is necessary to determine whether there are bubbles or other optical flaws in the rod or on the surfaces. A He-Ne laser beam which has been spatially filtered and expanded to a 10 mm beam is very useful for such inspections and can also be used in the alignment apparatus.

The laser mirrors are also critical components of the laser cavity. These mirrors are dielectric-coated quartz or laser glass substrates which are flat to one-tenth wave. Many types of mirrors are available from suppliers such as Valpey, Spectrum Systems, and Laser Energy. The reflectivity and wavelength, as well as the diameter, focal length, and wedge angle (if any), have to be specified for laser mirrors. The combination most commonly used is a rear reflector (99% reflectivity) which is slightly focusing, with a radius of curvature of 5-10 m, and a flat wedged mirror for the front (output) element. This makes a more stable laser cavity than one in which both elements are flat [10]. The reflectivity of the front mirror is a matter of choice. If one wants to obtain consistent mode-locked behavior with long trains of pulses (of the order of fifty pulses in a train) and at somewhat reduced power (0.1-1 GW), then the reflectivity should be in the range from 60 to 75%. If one wants to obtain shorter trains with higher powers (1-10 GW) and at some sacrifice in consistent mode-locked behavior, then the choice should be in the range from 50 to 30% reflectivity. The wedged front mirror means that no reflection from the second surface of the mirror will be fed back into the laser cavity, so that only a single picopulse will be circulating in the cavity. If a mirror with parallel surfaces is used as the output element, it will form a small cavity within the laser cavity and there will be 5-10 picopulses separated by the one-way trip time in the mirror circulating in the cavity.

Many different dye-cell configurations have been used. The
simplest choice is a high optical quality glass cell (glass seems to
withstand high optical powers as well as, if not better than, quartz)
of path length of 1-2 mm. No special coating is necessary if the
cell is placed at the Brewster angle with respect to the laser beam.
A dye cell with coated windows is available from Eastman Kodak. The
difficulty with a simple cell such as this is that there will inevitably be a subsidiary or satellite pulse formed within the laser cavity. The satellite pulse will grow to the same intensity as the main
pulse as the train develops. This pulse corresponds to a light fluctuation which coincides with the principal pulse at the dye cell, but
which is traveling in the opposite direction. This may be easily
recognized by the fact that the spacing between the pulses corresponds
to the distance between the dye cell and the rear reflector. The
satellite pulse may be suppressed by either incorporating the rear
reflector into the dye cell or using more than one dye cell in the
laser cavity. Longer path-length dye cells also tend to suppress
the satellite pulse, but they are not desirable because of dispersion
effects and the possibility of self-focusing in the solvent. Because
of the high light intensities within the laser beam, microscopic impurities in the dye solution are burned and form char which tends to
absorb and scatter the laser light. A recirculating flow system has
been used to continuously filter and replenish the dye cell. Considerable care should be exercised as to the cleanliness of the cell
and the dye solution. A 1 μ pore-size filter inert to dichloroethane
is available from Millipore.

Two dyes for mode-locking a neodymium:glass laser are available
in dichloroethane solution from Eastman Kodak, A9860 and A9740. Both
of these dyes are polymethine chains which are linked at each end
either to a thiocarbocyanine residue or a carbocyanine residue. The
structure of these dyes has been deduced by Pastor, Kimura, and
Soffer [11]. Both of these dyes are extremely sensitive to traces
of acid in the dichloroethane. If the optical density decreases rapidly, then activated alumina should be used to remove the traces of
acid in the dicloroethane. Both of these dyes are also thermally

unstable and should be stored in a refrigerator or freezer. They
inevitably decompose in the laser because of stray flashlamp light
and side reactions which occur under irradiation with 1.06 μ light.
the mode-locking dye A9860 is generally preferred because of its
better spectral match to the laser, the shorter pulses which are
produced, and its better stability relative to A9740.

The dye most commonly used for mode-locking ruby lasers is
1,1'-diethyl-2,2'dicarbocyanine iodide, which is also available from
Eastman Kodak. This dye is very unstable to light and oxygen. It
is soluble in methanol and water, both of which have been used for
mode-locking with this dye.

After obtaining the high-quality optical components for the
laser, considerable thought and care should go into assembling the
components into a system. It is important that the alignment of the
front and rear reflectors be accurate to within 5 seconds of arc and
that the alignment be stable. Another consideration is that the optical axis of the laser mirrors be coincident with the axis of the
laser rod. With a laser rod whose ends are cut at the Brewster angle,
this entails not only the orientation of the rod in space but also
the rotation of the faces of the rod about the axis. The most direct
method of accomplishing these ends is to use an autocollimating
alignment telescope. This device is used to center the mirrors and
the faces of the rod on an optical axis; then in the autocollimating
mode it is used to align the reflecting surfaces of the mirrors so
that they are perpendicular to the optical axis within a few seconds
of arc. With a curved rear reflector in the laser, it is necessary
to use an autocollimating alignment telescope in order to bring the
reflection of the curved surface into focus. The usual fixed auto-
collimators which are permanently focused at infinity can only be
used to approximate the true alignment of the curved surface.

Since the light sources used with an autocollimating alignment
telescope do not have sufficient intensity for easy visibility of
the light reflected from the front reflector in the autocollimating
mode in the configuration shown schematically in Fig. 4, it may be
necessary to use a special high-intensity source or to use an

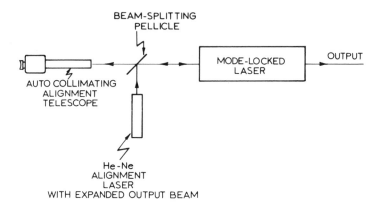

FIG. 4. Optical setup for the alignment of a mode-locked laser.
The axes of the deflected He-Ne laser beam and the autocollimating
alignment telescope are coincident. The optical axis of the mode-
locked laser is then aligned along this axis. The reflected light
from the laser mirrors enters the telescope, which is used to align
the mirrors to within 5 seconds of arc. If a curved rear reflector
is used in the laser, the telescope is adjusted to bring the reflection from this surface into focus. It is then readjusted to bring
the reflection from the flat front mirror into focus.

expanded He-Ne laser beam in order to see the reflection. An expanded

He-Ne laser beam is useful not only as an auxiliary to the telescope

but is also as an aid to aligning the optics of the experiment, since

it traces out the path of the mode-locked laser beam (assuming that

the He-Ne laser beam enters the pulsed laser through the rear reflec-

tor).

In the absence of an autocollimating alignment telescope, other

techniques can be used. The simplest is to use a He-Ne laser to

trace out the optical path and to physically center the mirrors and

the faces of the laser rod on the beam. The alignment can be achieved

by reflecting the He-Ne laser beam back onto itself via a long optical

path (more than 20 ft). If the alignment is sufficiently close, then

the reflections of the mirrors will show the Fabry-Perot interference

fringes. If the fringe pattern is centered on the reflections (this

is usually an ambiguous experiment), then this is the best that can

be done with such a system. It is usually good within 15 seconds of
arc, depending upon the skill of the individual. Another, more sophisticated technique for the final alignment has been described by
Fisher [12]. He utilizes a special infrared stimulateable phosphor
and split-image photography to record the line of lasing within the
laser rod. This is an empirical method to determine that the optical
axis of the laser mirrors is coincident with the axis of the laser
rod. Another technique is to record the far-field burn pattern on
a Polaroid print with solvent in the dye cell [13].

Assuming the laser is properly aligned, the next step is to determine the appropriate dye-cell transmission and optical pumping
level for best mode-locked behavior. The transmission of the dye is
measured at low light levels with a spectrophotometer at the wavelength of lasing, either 694 nm or 1.06 μ. Typical values range from
60 to 70% dye transmission for front mirrors with 30 to 45% reflectivity, and from 70 to 90% dye transmission for front mirrors with
50 to 85% reflectivity. In the former case, the pulse trains tend
to be shorter, with powers ranging from 1 to 10 GW; in the latter
case, the pulse trains are longer (50-75 pulses) with power ranging
from 0.1 to 1 GW. These numbers are necessarily approximate, since
each setup will differ in details and the pulse trains are not reproducible over long periods of time.

The optical gain and energy storage in the rod is determined by
the amount of light emitted by the flashlamp, which is a function of
the energy stored in the capacitor. This parameter is usually varied
by changing the voltage to which the capacitor is charged. The most
consistent results are usually obtained by operating slightly above
the lasing threshold. Higher voltages (and thus higher gain and
energy storage by the rod) result in multiple pulsing, that is, more
than one pulse train will be emitted by the laser in a given "shot."
Usually the second of the pulse trains is poorly mode-locked, and
this can lead to extraneous results.

During and after the laser "shot," the laser rod becomes
optically distorted because of thermal gradients created within the

rod by the discharge of the flashlamp [14]. This phenomenon can be
easily observed by passing a He-Ne laser beam through the laser rod
and noting the distortions which appear immediately upon discharging
the flashlamp. It usually takes one or two minutes for the rod to
return to thermal equilibrium. This, then, is the rate-determining
step in establishing how often the laser may be fired with consistent
mode-locked behavior.

There have been reports [12,15] of longitudinally mode-locked
lasers which have been specially equipped to oscillate in the TEM_{00}
transverse mode. These lasers have a diffraction-limited output
with a Gaussian intensity distribution across the beam profile. This
is useful with laser amplifiers and experiments requiring nearly
plane wavefronts. The simplest technique is to place a small aperture (6 mm or less) within the laser cavity. This produces a small-
diameter TEM_{00} output beam and a higher lasing threshold. The other
technique uses two 50 cm focal-length lenses which are separated by
1 m and placed between the laser rod and the output mirror. Midway
between the two lenses is a 0.1 mm diam diamond pinhole, accurately
positioned (vertically and horizontally) at the exact (common) focal
point of the two lenses. This apparatus effectively filters the
laser light so that only the TEM_{00} mode will propagate through it.

III. PICOSECOND PULSE MEASUREMENTS

In order to determine whether the laser is mode-locked, that is,
whether it is emitting a periodic train of picosecond pulses, it is
necessary to have a photodetector and oscilloscope of sufficient
real-time bandwidth to record the individual pulses in the train.
Let us suppose that the optical distance between the front and rear
reflectors of the laser is 1.5 m. The round trip time is 10 nsec.

No commercially available oscilloscope has sufficient real-time bandwidth (or sufficiently short risetime) to resolve the picosecond duration of each light pulse (as of this writing).[*]

The Tektronix Model 519 has a 1 GHz real-time bandwidth (or risetime of 0.3 nsec). This oscilloscope in combination with a fast photodetector is found in nearly all laboratories using mode-locked lasers. This oscilloscope displays the individual pulses as peaks 1 nsec wide on the screen. Other oscilloscopes having somewhat less bandwidth (typically 500 MHz) may also be used. These oscilloscopes are more versatile for general laboratory use. Since the oscilloscope trace vanishes from the screen very quickly, it is necessary to record the trace with an oscilloscope camera of the highest writing speed, i.e., a large relative aperture. A typical trace is shown in Fig. 5. Each spike in the photograph represents the response of the oscilloscope and photodiode combination to the picosecond light pulse. The height of each spike is proportional to the total energy (not the power) of the picopulse.

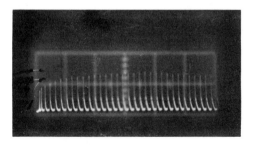

FIG. 5. A mode-locked train of picopulses as seen on a Tektronix Model 519 oscilloscope. This sort of photograph is used to diagnose the mode-locking of a laser, but does not indicate the picosecond duration of the light pulses.

---

[*]Sampling oscilloscopes do have risetimes of the order of picoseconds, but they cannot be used with pulsed lasers of this sort for two reasons: irreproducible pulses are emitted from one shot to the next, and the time between shots is too long for them to operate effectively.

There are two requirements for the photodetector: it must
respond to the infrared light of the ruby or neodymium:glass laser,
and it must have a risetime less than or equal to that of the oscilloscope. Photodetectors of this sort are available from a variety
of manufacturers at varying cost. They range from the large (and
expensive) planar vacuum photodiode types which operate with a bias
voltage of 2 kV or more, to the small solid-state PIN photodiodes
which operate with reverse bias voltages of less than 50 V. In this
latter category is the photodiode mount and circuit shown in Fig. 6.
Only one machined part and two commercially available components are
necessary to build this inexpensive photodetector suitable for use
with mode-locked lasers. Its spectral response ranges from 400 nm
to 1.1 μ and the inherent risetime of the photodiode is less than a
nanosecond. The mount is easily assembled with a filter holder, and
a small battery (37½ V) may be used to supply the bias voltage.

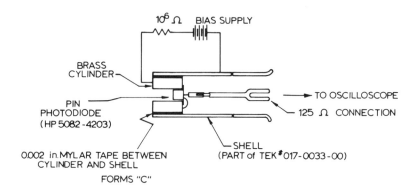

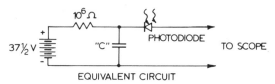

FIG. 6. Construction of a photodetector mount suitable for use
with mode-locked lasers. A photodiode (HP 5082-4203) and panel con-
nector (Tektronix Part No. 017-0033-00) together with a brass cylin-
der which is biased by a small battery (37.5 V) make up the photodetector. This is easily combined with a filter holder to make a
convenient laboratory instrument.

Photographs of the oscilloscope traces such as shown in Fig. 5 are useful diagnostics for determining whether the laser is well mode-locked. If the operating voltage is too high, multiple traces will be observed, corresponding to more than one train of pulses in a given "shot." If the dye solution is either too concentrated or too dilute, a broad envelope will be observed with few or no spikes. This is the behavior of a Q-switched laser. The only method to determine the best dye concentration for mode-locked pulses is to systematically study the pulses over a range of concentrations.

These photographs of the oscilloscope trace indicate the development of the train of pulses, but do not indicate the true duration of the picopulses. In order to determine the duration of the picopulses, we must make the light pulses photograph themselves. Several techniques have been developed to accomplish the measurement of the pico-pulse duration [9,16]. All of these techniques utilize one or another of the nonlinear interactions of light with molecules. These non-linear effects are easily observed with the gigawatt power levels of picopulses. The first and most popular method developed is the two-photon excitation of fluorescence technique, commonly called the two-photon fluorescence or TPF method.

Ordinary absorption of light by a molecule corresponds to a transition between the ground state and an excited state of the opposite parity caused by the absorption of a single photon of the appropriate wavelength. In 1933 Goeppert-Mayer [17] predicted that two-photon absorption was also possible. If two photons were simultaneously absorbed by an atom or molecule, then a transition would be observed between the ground state and an excited state of the same parity and an energy corresponding to the sum of the energies of the two photons which were absorbed, as shown schematically in Fig. 7. There is no real intermediate state; the molecule does not absorb light at the intermediate wavelength. Two-photon absorption is not ordinarily observed because of the limited light intensities which can be achieved with classical light sources. In 1961 Kaiser and Garrett [18] first demonstrated two-photon absorption experimentally in a europium-doped crystal with a ruby laser. The probability

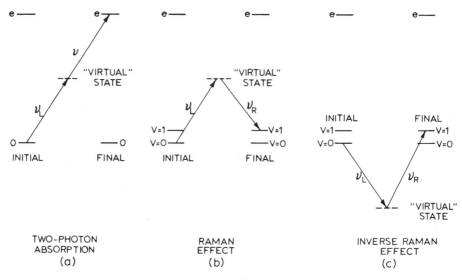

FIG. 7.   Schematic diagrams of (a) two-photon fluorescence, (b) the Raman effect, and (c) the inverse Raman effect. All of these processes involve the "virtual" intermediate state, which can be thought of as the sum of all the excited states of the molecule.

of one-photon absorption is proportional to the intensity of the incident light, but the probability of absorption for a two-photon transition is proportional to the product of the light intensities of each of the photons to be absorbed. If two identical photons are absorbed, then the probability of a two-photon transition is proportional to the square of the incident light intensity.

If the atom or molecule makes a two-photon transition to an excited state, then fluorescence takes place in the usual way (taking into account the fact that the parity of the excited state reached in this manner is the same as that of the ground state). This then is a nonlinear process, namely, that the intensity of the fluorescent light is proportional to the square of the intensity of the exciting light.

A schematic diagram for the two-photon fluorescence technique is shown in Fig. 8. The laser beam is split into two parts, focused to beam diameters of 2-3 mm, and recombined in a cell containing a solution of a highly fluorescent dye which does not absorb at the laser

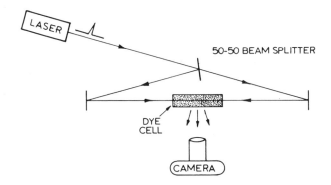

FIG. 8. Optical arrangement of a two-photon fluorescence measurement of picopulse duration. The pulses collide in the center of the cell and cause an increase in the probability for two-photon absorption, and consequently an increase in the fluorescence intensity.

wavelength (twice the energy). A camera records the path of the laser beam by photographing the fluorescent light of the dye (which has been excited by two-photon absorption). At the point in space where the picosecond pulses collide, there will be an increase in the two-photon absorption because of the increased intensity of light at that point. The fluorescence intensity at the point of collision is three times as great as the intensity of fluorescence at other points along the path of the laser beam.* Imperfect mode-locking yields contrast ratios less than 3:1. An example of two-photon fluorescense is shown in Fig. 9. The time duration of the 1.06 μ picosecond pulse is obtained by simply measuring the length of the bright spot of yellow fluorescence of rhodamine 6G. Typical pulse lengths for neodymium:glass lasers are 2-6 psec, while typical values for ruby lasers are 6-12 psec.

---

*One might expect that the ratio should be 4:1, but the colliding pulses are not in phase relative to one another over macroscopic distances so that an average of all possible relative phase angles must be taken in account.

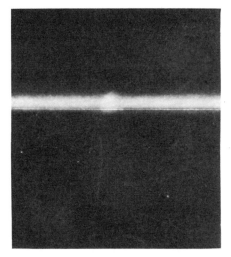

FIG. 9. Photograph of a two-photon fluorescence measurement.
For perfect mode-locking the ratio of the fluorescence intensities
observed at the point of collision and the remainder of the trace
will be 3:1. The duration of the picopulses is obtained by measuring
the length of the increased fluorescence. This method will not detect any asymmetries in the shape of the picopulse.

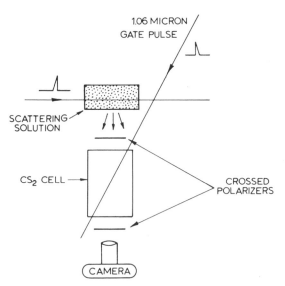

FIG. 10. Picopulses can be measured in flight by using a camera
with a very fast shutter. In this technique the 1.06 μ picopulses
operate the traveling-wave Kerry-cell shutter.

A unique experiment performed by Duguay and Hansen [19] is to literally photograph the visible 530 nm picopulses in flight. This is shown schematically in Fig. 10. The infrared 1.06 μ light pulses drive a fast shutter to freeze the green pulses in flight. The principles of this experiment are similar to those of the time-resolved emission technique which is explained below.

IV. NONLINEAR OPTICS

If the only available wavelengths of picosecond pulses were the fundamental wavelengths of the ruby (694 nm) and neodymium:glass (1.06 μ) laser, then they would be of very limited usefulness for experiments in photochemical kinetics. However, the extremely high powers of these light pulses enable one to exploit nonlinear effects by which one may change the wavelength of these pulses to nearly any desired wavelength in the near infrared, visible, or near ultraviolet. The techniques include second- third-, and fourth-harmonic generation, transient stimulated Raman scattering, stimulated fluorescence, and self-phase modulation.

The polarization of the electrons in a medium by the electric vector of a light wave gives rise to a second light wave which interferes with the incident light in such a way as to bring about such familiar phenomena as the refraction and reflection of light at an interface and the reduced velocity of light in material media. The polarization wave is the response of the electrons of the crystal or molecule to the applied force of the electric field of the light wave. In general only that part of the polarization which is linear in the applied electric field is observed because the force exerted by the light wave upon the electron is small compared to the forces binding the electron in the crystal. At sufficiently high light intensities, however, the force of the light wave can become comparable to the crystal fields, and the polarization will show a nonlinear response. The first nonlinear term is proportional to the square of the electric

vector of the incident light. It can be readily seen that this term
will cause two effects, namely, a dc polarization of the crystal and a
polarization at twice the frequency of the incident light. The situation is analogous to that of an electrical diode or rectifier. The
applied force is the voltage and the response is the current. If one
applies an alternating voltage to a diode, there will result a current
which can be resolved into a dc term and a term which is at twice the
frequency of the applied voltage.

Second-harmonic generation with picopulses is achieved by shining
the laser light beam through a crystal of potassium dihydrogen phos-
phate (KDP) at a specific angle with respect to the optic axis of
this uniaxial noncentrosymmetric material. When the laser light
travels through the crystal as an ordinary ray at this angle, called
the phase-matching angle, the second-harmonic (twice the frequency or
one-half the wavelength) light which is generated by the nonlinear
polarization travels as an extraordinary ray at the same phase velocity. This means that the two light waves, the fundamental and second
harmonic, are exactly in phase and coincident in time and space
throughout the length of the KDP crystal. Furthermore, they are
polarized orthogonally to one another. With 1.06 μ picopulses it is
a fortunate circumstance that in KDP not only do the phase velocities
of the two waves match at this angle but the group velocities are
also the same. This means that there is no broadening of the 530 nm
picopulse because of dispersion mismatch relative to the 1.06 μ picopulse. This simultaneous phase match and dispersion match does not
occur in other crystals used for second-harmonic generation.

For optimum second-harmonic conversion the KDP must be aligned
relative to the laser beam to within 10 seconds of arc. It is not
economically feasible to cut and polish the faces of the crystal more
precisely than to within 5 minutes of the exact angle for phase matching. The final alignment of KDP is done in situ by systematically
observing the conversion efficiency versus orientation. KDP is a
hydroscopic material. In order to protect the crystal and minimize
reflection losses, the crystal can be sealed in a cell with an index-
matching fluid or mineral oil. KDP crystals can be obtained from

Isomet and other manufacturers. The usual crystal sizes range from
12 to 24 mm. Typical conversion efficiencies range from 5 to 20%
for mode-locked ruby and neodymium:glass lasers. Higher conversion
efficiencies from the fundamental to the second harmonic are achieved
in specially designed TEM_{00} mode Q-switched oscillator-amplifier combinations.

A photon can be scattered by a molecule with an (instantaneous)
change in frequency or energy. The photon energy is reduced by an
amount equal to the quantum increase in vibrational energy of the
molecule. This scattering phenomena is called the Raman effect and
is shown schematically in Fig. 7(b). At low light intensities a
small but measurable number of photons are Raman scattered. At the
high light intensities, which can be achieved by focusing picopulses
into a suitable liquid such as methanol, water, or benzene, conversion from the incident to the Raman-scattered light occurs with high
efficiency. This is called stimulated Raman scattering. For example,
up to 80% peak conversion from 530 nm picopulses to 627 nm picopulses
has been observed in ethanol [20]. It has also been shown that the
stimulated Raman scattered picopulses are of shorter duration than
the initiating or pumping picopulses [21]. Colles has demonstrated
that pulses as short as 0.3 psec can be generated using a 530 nm
picopulse train to pump a stimulated Raman oscillator [22]. Several
materials are available for stimulated Raman scattering, including
acetone, trichloroethane, calcite, methane, and sulfur hexafluoride
(at high pressure) [23].

In addition to stimulated Raman scattering there is stimulated
fluorescence emission from certain dyes. The difference between the
phenomena is that in fluorescence the photons cause a transition to
a real excited state which has a finite lifetime, whereas in Raman
scattering there is no real intermediate state: the incident photon
instantaneously (within the limits of the uncertainty principle)
undergoes a change in direction and frequency. Stimulated fluorescence has been demonstrated in rhodamine 6G, rhodamine B, acridine
red, and several polymethine cyanine dyes [24-26]. The wavelength
and duration of the picopulses which are emitted by the stimulated

fluorescence technique vary with concentration, pumping powers, path length, and solvent. Typically these pulses are longer than the pumping pulses, and the wavelengths for a dye like rhodamine 6G will vary from 560 to 575 nm. Conversion efficiencies are far less than for Raman scattering, and will be in the range from less than 1 to 5%.

Recently Alfano and Shapiro discovered that when 530 nm pico-pulses are focused into liquid argon and krypton, the output contained picopulses which were spectrally broadened into a continuum ranging from 300 to 700 nm [27]. They also demonstrated that this same broadening could occur in sodium chloride, in certain glasses, and in quartz and calcite crystals [28]. They attribute the effect to the severe distortion of the electronic motions in the presence of the intense optical electric fields at the focus of the lens. If the electric fields of the light are comparable to the field from the nucleus, then the response of the electronic motion cannot be approximated by linear, quadratic or even cubic terms, but in fact will be a complex nonlinear function. The nonlinear response of the electron is the cause of the broadening of about 1% of the incident light into a "white" picopulse. Alfano and Shapiro call this effect "self-phase modulation." White picosecond pulses can be used to explore such areas as the inverse Raman effect [29], which is shown schematically in Fig. 7(c).

Far-infrared (2-20 cm^{-1}) picopulses can also be generated by the optical rectification of 1.06 μ picopulses in a nonlinear medium such as LiNbO$_3$ [30]. The crystal rectifies the alternating fields so that one is left with the envelope of the picopulse. One may consider that the envelope of the picopulse is simply the sum of several wavelengths in the far infrared. This then is a unique source of high-power (∼1 kW) short-duration infrared light.

V. PICOSECOND SPECTROSCOPY

Both emission and absorption spectra can be observed with pico-second time resolution using a mode-locked laser in conjunction with

relatively simple optical apparatus and a spectrograph. In the arrangements which are described below it is possible to determine both
the wavelength and time dependence in a single laser shot. In a
typical experiment picopulses at one or more wavelengths are various-
ly used to excite or interrogate the sample and to operate a light
gate (in an emission experiment). Essential to the success of these
methods is the short time duration and the high power of the fundamental picopulses. The high powers available enable one to derive
pulses of appropriate wavelengths synchronized with one another and
of sufficient energy to create an observable number of excited states.

At low light intensities (less than 1 kW/cm^2) the optical density or transmission of a dye is independent of light intensity, since
the rapid rate of relaxation from the excited state to the ground
state ensures that the ground state is not depleted. At the higher
light intensities that are available from Q-switched or mode-locked
lasers, the ground-state population may be significantly depleted.
Assuming that the excited-state molecules do not absorb light at
the same wavelength, the depletion of the ground state implies a decrease in the optical density, i.e., the transmission will increase.
This is the principle of the passive-dye Q-switch which is used to
mode-lock ruby and neodymium:glass lasers.

The light intensity at which the dye becomes significantly
bleached is inversely proportional to the product of the absorption
cross-section and the relaxation time. With the advent of picosecond
pulses it was possible to determine directly the relaxation time by
using a strong pulse to bleach the dye and a weaker pulse to interrogate the transmission of the dye at subsequent times. Scarlet,
Figueira, and Mahr [31] used an arrangement similar to that of
Shelton and Armstrong [32], shown schematically in Fig. 11. In this
experiment a 1.06 μ pulse incident from the left bleaches the Kodak
9860 dye and an attenuated pulse reflected from the right interrogates the transmission. Note that this technique utilizes the per-
iodicity of the train. The interrogating pulse is the member of
the train which precedes the bleaching pulse. In order to obtain
the entire decay curve for the Kodak 9860 dye it is necessary to

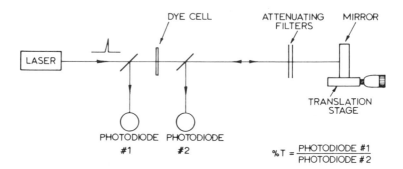

FIG. 11. The relaxation time of a dye is measured with the collinear technique shown. A strong pulse incident from the left bleaches the dye, and a weak pulse reflected from the right interrogates the transmission. The distance from the cell to the reflecting mirror is adjusted so that the interrogating pulse arrives shortly before, during, or after the bleaching pulse. Many laser shots are required to obtain the complete time dependence of the relaxation time.

take many laser shots with different delays between the bleaching and interrogating pulses. A similar technique has been used by Eisenthal and Drexhage [33] to obtain information about the rotational relaxation time. In this case the plane of polarization of the interrogating beam is rotated 45° with respect to the plane of polarization of the bleaching pulse. The components of the interrogating pulse perpendicular and parallel to the bleaching pulse are separately and simultaneously analyzed for dichroism induced by the bleaching pulse (which preferentially excites those molecules whose transition moment is parallel to the plane of polarization).

The collinear techniques described above offer high sensitivity and time resolution, but suffer from the fact that many laser shots are required to obtain even a single datum point. Since the output of mode-locked lasers of this sort is not reproducible from shot to shot, and since there is the possibility of damage to the laser components or optics, a technique wherein all the information is collected in a single laser shot is a desirable goal.

We have successfully used a simple technique for the observation of the relaxation of a dye from the excited state using a single

laser shot [34]. The technique uses two picosecond light pulses
(which in general may or may not be of the same frequency). During
the passage of an intense bleaching pulse through a dye-containing
sample cell, the cell is uniformly illuminated by a second interrogating pulse traveling at right angles to the first, as shown in
Fig. 12. The short duration of the picopulses enables us to "freeze"
the bleaching pulse within the cell and photograph the recovery process. From the transparency of the dye, measured along the path of
the bleaching pulse, we obtain the complete time history of the dye,
before, during, and after the passage of the intense bleaching pulse.
This technique is applicable to both fluorescent and nonfluorescing
species, the only requirement being that the absorption band is saturable by the picosecond light pulse. Generally, this means that only
absorption bands with extinction coefficients greater than 10^3 liter/
mole-cm are suitable for this type of experiment. With bands having

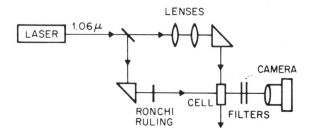

FIG. 12. The cross-beam technique enables the complete time
history of the bleaching and relaxation process to be taken in a
single laser shot. Collimating lenses focus 90% of the mode-locked
laser output to a 2 mm diam bleaching light pulse, which traverses
the rectangular cell along the 1 cm path. A beam splitter diverts
10% of the output to form the weak interrogating pulse, which traverses the rectangular cell along the 2 mm path at right angles to
the bleaching pulse. Suitable filters are used so that only the
1.06 μ light enters the camera. The short duration of each of these
pulses "freezes" the bleaching light pulse within the cell. The
light entering the camera is a record of the density of the dye as
a function of time.

smaller extinction coefficients, the photons in the picopulse (approximately 10^{15}) will be absorbed before there is any appreciable fractional change in the ground-state population. Given the proper circumstances, this method is also suitable for excited-state absorption studies and two-photon fluorescence experiments.

In this experiment the output of the laser is collimated to a 2 mm diam beam by means of an inverted telescope, namely, two lenses which are separated by a distance equal to the sum of their focal lengths. For the best results these lenses should be well aligned and centered along the optic axis and the distance between the lenses should be adjusted empirically, since the focal lengths for visible wavelengths is less than that for the near infrared. The ratio of the focal lengths determines the ratio of the input and output beam diameters. A pellicle splits off 8% of the 1.06 μ picopulse which forms the broad weak beam of interrogating light which illuminates the dual path length cell (2 x 10 mm). Color filters (Schott RG-9) and a narrow-band (10 nm) interference filter are placed in front of the camera so that only 1.06 μ light is recorded on the infrared-sensitive film (Polaroid Type 413). If these filters are not used, stray flashlamp light will enter the camera and obscure the results. The implicit assumption upon which this experiment rests is that the interrogating picopulse is of uniform intensity across the width of the cell, so that the differences in light intensity reaching the film arise solely out of the bleaching of the dye. In order to minimize the spurious effects caused by nonuniformities in the interogating beam, the dye cell and camera are placed some distance from the output of the laser (usually more than 2 m) and the camera is focused on the dye cell. An additional device which was used was a coarse transmission diffraction grating (20 lines/mm) placed 1 m away from the cell. This grating diffracts the incoming beam very slightly, so that the cell is contained within the central maximum. A better way to accomplish this task is to use a 1:1 telescope with a spatial filter to eliminate the nonuniformities. Yet another solution is to use a TEM_{00} single transverse mode mode-locked laser.

The data from experiments with Kodak Q-switch dye 9860 are shown in Fig. 13. The concentration of 9860 used in these experiments was adjusted so that the optical density of the cell was equal to one unit along the 2 mm path, and therefore was equal to 5 units of optical density in the 10 mm direction. This concentration was used in order to obtain the necessary contrast between the bleached and unbleached portions of the dye. Higher concentrations of dye are not desirable because of the residual absorption of the bleaching 1.06 μ picopulse. The timing of the two pulses is crucial to the success of this experiment. Every effort was made to keep the paths outlined by the two pulses exactly rectangular and to compensate for the differences in path lengths introduced by lenses and filters. In order to show that the effects observed were genuine, various thicknesses of quartz were placed in the interrogating beam so that it was delayed relative to the arrival of the bleaching pulse. The bleached portions of the dye were observed to move forward in the cell the proper distance corresponding to the delay which was introduced.

Because the dye cell is 2 mm thick, the bleaching pulse moves while the cell is interrogated. Thinner cells would have minimized this effect, but also would have decreased the optical density (for the same dye concentration) and lessened the contrast ratio between the bleached and unbleached portions of the dye. The observed transparency is of course a convolution of the two picopulses and the saturation recovery time of the dye. In addition, filaments of approximately 50 μ diam were visible in these photographs. These are high-power portions, i.e., "hot spots," of the laser output, which bleach the dye locally.

In order to analyze the data, a simple model for the process was derived. Since the observed pulse lengths of the picopulses were short compared to dye relaxation time and the attenuation of the bleaching light pulse was negligible over a distance of 2 mm, we could simply assume that the picopulses were "delta functions" for this experiment. At any point in the cell the absorbance α, instan-

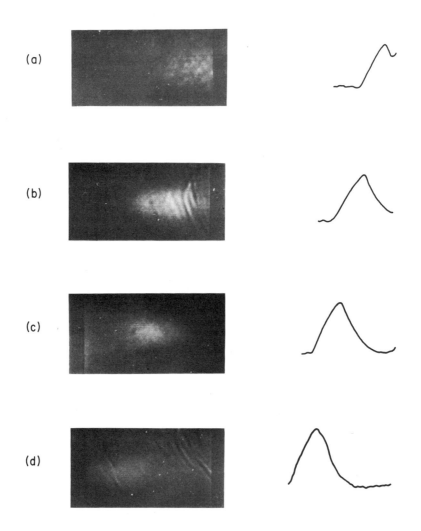

FIG. 13. Photographs of Kodak Q-switch dye 9860 (OD = 5/cm at 1.06 μ). The bleaching light pulse travels from right to left in these pictures. Various thicknesses of quartz inserted into the interrogating beam delay the arrival time of the interrogating pulse relative to the bleaching pulse, thus moving the bleached portions of the dye to the left. The observed transparency of the dye is a convolution of the two light pulses and the saturation recovery time of the dye. Assuming a simple model for the dye saturation, we can measure the saturation recovery time from the trailing edge of the bleaching light pulse. We find a value of 9 psec for the Kodak Q-switch dye 9860.

taneously vanished upon the arrival of the bleaching picopulse and
recovered with a relaxation time τ, so that

$$\alpha(t) = \alpha_0 \qquad\qquad\qquad t < 0$$
$$\quad = \alpha_0 (1 - e^{-t/\tau}) \quad t \geq 0$$

where $t = 0$ is the arrival time of the bleaching picopulse, and α_0
is the unbleached optical density. When we consider that both the
bleaching and interrogating picopulses travel at the same velocity
through the solution, then we see that there are three regions in
the dye cell, the portion ahead of the bleaching pulse, the portion
of overlap between the interrogating and bleaching pulses (which is
equal to the thickness of the cell), and the portion after the
bleaching picopulse where the dye is recovering or relaxing back to
the ground state.

Using the approximation that both the interrogating and bleaching pulses are plane waves of negligible extent (i.e., delta functions in time) and considering the geometry of the cell, we arrive
at the following expressions for the optical density $\alpha(x)$ as a function of the distance, x, across the cell as it appears in the camera:

$$\alpha(x) = \alpha_0 \qquad\qquad\qquad\qquad\qquad\qquad\qquad x < 0$$
$$\quad = \alpha_0 \, [1 - (\delta/d)(1 - e^{-x/\delta})] \qquad 0 \leq x < d$$
$$\quad = \alpha_0 \, [1 + (\delta/d)(1 - e^{d/\delta}) \, e^{-x/\delta}] \quad d \leq x$$

In these expressions δ is the relaxation "length" of the dye,
that is, $\delta = v\tau$ where v is the group velocity of the picopulse. The
positions $x < 0$ represent the portion of the cell ahead of the
bleaching pulse, while $0 \leq x < d$ is the portion where the bleaching
and interrogating picopulses overlap. The length, d is equal to the
thickness of the cell (2 mm) in the direction of the interrogating
pulse. The transmission increases, i.e., the optical density decreases, over this length and is a good indicator of the validity
of the approximations used to derive these expressions. The finite
duration of the picopulses will tend to increase the length of this

portion of the bleached area. Although the data are somewhat noisy,
the length of the rising portion does not exceed 2.3 mm. The positions d $\leq$ x represent the portion of the dye after the bleaching
pulse has passed and the dye is relaxing back to the ground state,
so that the optical density increases with a relaxation "length" of
approximately 1.8 mm corresponding to $\tau \approx$ 9 psec. Note that both
the decreasing and increasing portions of the optical density have
the same relaxation length, so that the internal consistency of these
approximations can be checked. Similar measurements of the Kodak Q-
switch dye 9740 reveal similar results, $\tau \approx$ 10 psec.

These values are in close agreement with those obtained by
Mauer (9 and 8 psec, respectively) using an indirect method [35].
Scarlet et al., reported a value of 6 psec for 9860 [31].

We have also performed measurements at 530 nm using a KDP crystal to generate the second harmonic of the 1.06 μ laser output.
Experiments on several dyes (rhodamine 6G, acridine red, erythrosine
red, rosaniline, acetamine red, and crocein scarlet) revealed that
the fluorescent dyes were readily bleached and possess relaxation
times of the order of nanoseconds. Bleaching of the nonfluorescent
dyes was not apparent. We speculate that there was insufficient
power at this wavelength to bleach these dyes, although we cannot rule
out other possibilities such as an extremely rapid relaxation time
(less than a few picoseconds) or that the cross-section was so small
that the 530 nm picopulse was completely absorbed before it reached
the center of the cell.

The difficulty with the collinear technique such as used by
Scarlet et al. (see Fig. 11) is that many laser shots are required
to obtain a single datum point. The advantage of the cross-beam
technique (see Fig. 12) is that all of the data are obtained in a
single laser shot and the results are easily interpreted; in particular it is easy to diagnose a "bad shot" in which the picosecond
pulses were not of sufficient intensity or were too long. The disadvantage of the cross-beam technique is that there has to be a compromise between the ultimate time resolution (which would require a
thin cell) and the necessary optical density change (which requires

a thick cell). The minimum optical density change which can be resolved with film such as Polaroid (which is somewhat noisy) is 0.1
od unit. This could be improved with a photodetector array in the
place of film.

A combination of the best features of each of these geometries
is obtained by using two nearly collinear beams and a step delay in
the interrogating beam as shown in Fig. 14. This technique is a
variant of that introduced by Topp for picosecond time-resolved emission studies [36]. The technique allows the strong bleaching beam to
travel almost parallel to the interrogating beam, which is divided
up into parallel sections with a step progression of time delays
which are introduced with a stack of microscope slides arranged in
step sequence. This geometry allows longer path lengths without

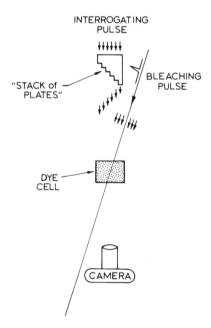

FIG. 14. The echelon or "stack of plates" technique. This
technique combines the advantages of the collinear and the cross-
beam techniques. Adjacent segments of the interrogating beam are
delayed by the various thicknesses of glass. This allows the transmission of the dye to be sampled at various times relative to the
bleaching pulse.

loss of time resolution, and the time differences between the various segments may be as small as one chooses. Each millimeter of
glass introduces approximately 1.67 psec of delay relative to the
same path length in air. Again in this technique it is assumed that
the cross section of the laser beam is of uniform intensity. This
can be accomplished with a TEM_{00} transverse mode laser as was
mentioned previously. There have not been any experiments published
utilizing this geometry for absorption studies so far as the author
is aware.

An excited-state absorption experiment which is unique in its
application was suggested by Longuet-Higgins and performed by
Rentzepis [37]. In this experiment a 530 nm picopulse is used to
excite azulene in solution to the highest vibrational levels of the
first excited singlet, S_1 (which does not fluoresce). A 1.06 μ
picopulse traveling in the opposite direction in the cell is absorbed by the molecules remaining in the highest virational levels
of S_1, promoting them to the second excited singlet S_2, which then
undergoes fluorescence. A camera photographs the fluorescence of
azulene in the cell along the path of the colliding picopulses. By
measuring the length of the azulene fluorescence track in the cell
one may determine the relaxation time of azulene from the highest
vibrational levels. Rentzepis has reported a value of 7.5 psec from
his experiments. Drent, Van der Deijl, and Zandstra have reported a
similar experiment using a mode-locked ruby laser [38]. In this case
the 694.3 nm picopulses excite azulene directly to the lowest vibrational level and a subsequent 694.3 nm picopulse is absorbed, raising
the molecule to the second excited singlet S_2, which then fluoresces.
They attribute the observed relaxation time of 4.0 psec entirely to
intersystem crossing, since they observe a heavy-atom effect with
CCl_4. They conclude that the measurements of Rentzepis might include
intersystem crossing as well as vibrational relaxation.

We have mentioned previously the use of transient stimulated
Raman scattering to produce picopulses of differing wavelengths.
The normal Raman effect involves the spontaneous emission of a pho-
ton of frequency which is lower than the incident photon frequency.

The inverse Raman effect is shown schematically in Fig. 7(c). An intense monochromatic light beam can stimulate absorption of light of frequencies greater than the frequency of the incident light. The molecule has a transitory energy deficit in the case of the inverse Raman effect, while there is a transitory energy surplus in the normal Raman effect. Energy need not be conserved for these intermediate "virtual" states, since they only have a transient (10^{-15} sec) existence and according to the uncertainty relation it is impossible to determine the energy of such a short-lived state with any precision. In order to observe the inverse Raman effect it is necessary to have intense monochromatic light and simultaneously a weak continuum. The effect has been observed by McLaren and Stoicheff [39] using a Q-switched ruby laser and the spontaneous fluorescence of rhodamine 6G excited by the second harmonic of the ruby laser. Alfano and Shapiro [29] demonstrated the inverse Raman effect with 530 nm picopulses and a broadband continuum provided by the self-phase-modulated "white" picopulse. They have observed four Raman lines in benzene which have not been previously reported. There is the possible application of this technique to observe the Raman spectra of excited-state molecules.

In order to obtain both the picosecond time- and wavelength-resolved absorption spectra of excited states, one might use the apparatus schematically indicated in Fig. 15. The 530 nm picopulses are used (or are wavelength shifted) to "pump" molecules of interest into the excited state, and the self-phase-modulated or "white" picopulse is used to interrogate the excited-state absorption. A spectrograph then records the absorption spectrum. The image on the focal plane of the spectrograph consists of light intensity versus wavelength (perpendicular to the slit) and versus time (parallel to the entrance slit). This information is easily recorded on film, or it might be recorded on an image intensifier or an array of detectors which could be read out electronically. The latter possibility offers more dynamic range and precision than is possible with film.

The Kerr cell is a familiar electro-optic device which consists of a pair of electrodes immersed in a polar liquid (such as nitro-

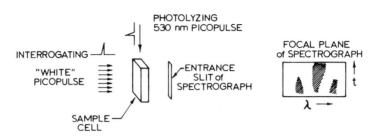

FIG. 15. The complete time and spectral resolution of an
absorption band may be determined with this picosecond version of
flash photolysis. The "white" picopulse interrogates the photolysis
cell and enters the spectrograph. The spectral dependence is displayed in the direction perpendicular to the entrance slit as usual,
and the time dependence is displayed in the direction parallel to
the entrance slit.

benzene). When a voltage is applied across the electrodes, the polar
molecules tend to align themselves along the electric field. The
effect of this (partial) alignment is to produce birefringence. In
1964 Mayer and Gires [40] showed that the powerful (polarized) light
pulses from a ruby laser would induce birefringence in liquids. The
optically induced birefringence arises out of the partial alignment
of molecules in the liquid. The anisotropic molecules experience a
torque which tends to rotate the axis with the greatest polarizabi-
lity into the plane of the electric vector of the light. In 1969
Shimizu and Stoicheff [41] studied the birefringence induced in
small-scale trapped filaments in CS_2 produced by 1.06 μ picopulses.
Duguay and Hansen [42] showed that the birefringence induced in CS_2
(not necessarily in filaments) by 1.06 μ picopulses could be used as
a traveling-wave Kerr cell. As the 1.06 μ picopulse travels through
the CS_2, it orients the molecules in its path; after the pulse has
passed, the molecules relax again to an isotropic distribution with
a characteristic rotational relaxation time of 1.8 psec. Duguay and
Hansen demonstrated the use of an ultrafast light gate with a nearly
collinear geometry as shown schematically in Fig. 16. As we noted
above in the discussion of relaxation times, the collinear geometry
offers better time resolution without sacrificing peak transmission,

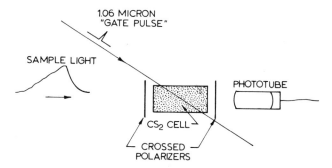

FIG. 16. The traveling-wave Kerr cell first demonstrated by Duguay and Hansen is shown schematically in this figure. The electric field of the picopulse induces birefringence in CS_2, which allows the weak light beam to pass through the cross polarizers momentarily. Many laser shots are required to obtain the complete time history of the emission.

but suffers from the fact that many laser shots are required even to obtain a single datum point. Since mode-locked lasers are unreliable and the reproducibility is less than perfect, a technique which uses a single laser shot to obtain all the data is desirable.

We demonstrated [24] a traveling-wave Kerr cell with a cross-beam geometry as shown in Fig. 17. The mode-locked neodymium:glass laser generates 1.06 μ picopulses which are polarized perpendicular to the plane of Fig. 17. These pulses are then collimated to 1.5 mm diam beam which travels through a 10 x 1 mm cell containing CS_2. The intense electric field of the 1.06 μ "gate" picopulse induces a segment of birefringence traveling at the speed of light with a 1.8 psec "tail." In this segment of the CS_2 the index of refraction perpendicular to the plane of Fig. 17 is slightly greater than that in the plane of the figure. Now let us assume that a weak light beam polarized at 45° to the plane of the figure and of uniform cross section is incident upon the cell in the direction perpendicular to the 1.06 μ pulse. Because of the birefringence induced in the CS_2, the plane of polarization of the weak light is slightly rotated and so it is partially transmitted through a second (crossed) polarizer. The net effect of the traveling-wave Kerr cell placed between

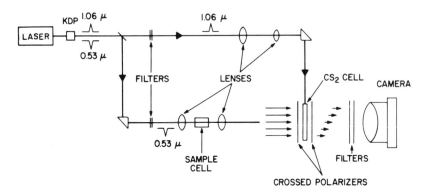

FIG. 17. The cross-beam light gate enables one to obtain the
time dependence of light emission in a single laser shot. Telescoped
1.06 µ picopulses travel along the 1 cm path of the CS$_2$ cell. The
electric field of the picopulse tends to orient the molecules parallel to the field. This gives rise to a transitory birefringence
that allows the sample beam to travel through the crossed polarizers
for several picoseconds. The effect is that of a shutter traveling
at the speed of light in front of the camera. If the light emission
is of uniform cross section, then the light intensity reaching the
film reflects the time dependence of the emission.

crossed polarizers is that of an opening 1 mm wide traveling at the
speed of light. The transmission of the shutter opening is proportional to $\sin^2(\phi)$, where ϕ is the angle of rotation. The angle ϕ is
related to the length ℓ of the birefringent medium (i.e., the interaction length), the optical Kerr constant of CS$_2$ (n_{2B}), and the
electric field of the gate pulse E, as follows:

$$\phi = \frac{2\pi}{\lambda} n_{2B} \ell E^2$$

If we place a camera behind this unique type of shutter and
assume that the incident light beam is of uniform intensity across
its cross section, then the light intensity that reaches the camera
across its field of view can be interpreted as the time history of
the incident light.

The characteristics of the cross-beam light gate were calibra-
ted with 530 nm picopulses generated by the 1.06 µ picopulses in KDP.

The time that the shutter is open is determined by the thickness of
the cell (1 mm = 5.3 psec), the rotational relaxation time of CS_2
(1.8 psec), and the duration of the 1.06 μ picopulse (2 psec). Experimentally we found that the apparent half-width of the "opening"
as recorded on a microdensitometer tracing, shown in Fig. 18(a), of
the Polaroid Type 47 print of the 530 nm picopulse corresponds to 7
psec, with rise and fall times (1/e) of about 2.5 psec. The peak
transmission of the shutter opening was of the order of 1%, and the
on-off ratio is greater than 500:1. Improved peak transmission and
greater on-off ratios can be obtained at the expense of time resolution by using a thicker CS_2 cell. Filaments were sometimes observed

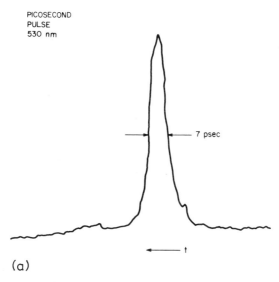

PICOSECOND
PULSE
530 nm

7 psec

t

(a)

FIG. 18. Microdensitometer traces of the Polaroid prints used
to record the light pulses. Since the CS_2 cell is 1 mm thick, the
apparent width of a light pulse will contain a contribution of 5.3
psec. In (a) is shown the trace of 530 nm pulses generated in KDP
by the 1.06 μ picopulse. Stimulated Raman scattering in ethanol is
shown in (b). This pulse is narrower in time than the 530 nm picopulse, but the differences are within experimental error. Stimulated emission in rhodamine 6G dissolved in methanol 5 x 10^{-5} M) is
shown in (c). The time duration of the latter pulse is approximately 6 psec.

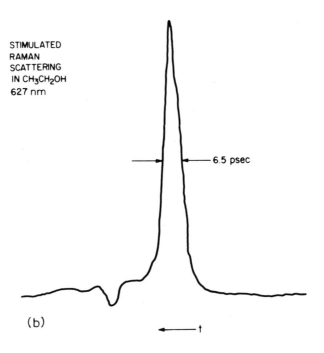

STIMULATED
RAMAN
SCATTERING
IN CH_3CH_2OH
627 nm

6.5 psec

(b)

t

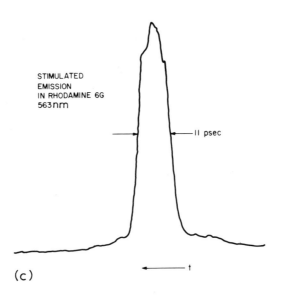

STIMULATED
EMISSION
IN RHODAMINE 6G
563nm

11 psec

(c)

t

in the cell, but these are thought to represent "hot spots" in the
laser output rather than self-focusing, since self-trapped filaments
in CS_2 are usually formed over longer path lengths.

Aside from the convenience of operation of the cross-beam geometry, there is also a built-in calibration of the picopulse lengths.
If the 1.06 μ and 530 nm picopulses are made to coincide at the center of the CS_2 cell, then a spot of light will appear at this point.
If, however, the pulses are longer than normal, i.e., a "bad shot,"
then light will appear on both sides of the central "coincidence"
point. As with the stimulated process, such as stimulated Raman
scattering and stimulated fluorescence emission, the light spot will
appear at the center point and will generally point toward the "exit"
of the 1.06 μ picopulses which correspond to later times. However,
if the 1.06 μ pulses are longer than normal, light will appear to
the right of the center point, toward the "entrance" of the gate
pulse.

Time-resolved stimulated Raman scattering in ethanol (λ = 627
nm, ν = 2928 cm^{-1}) is shown in Fig. 18(b). The apparent half-width
of the Raman light pulse is 6.5 psec, which is narrower than that of
the 530 nm pulse, but the differences are within experimental error.
Since there is always a contribution of 5.3 psec to the apparent
width due to the finite thickness of the cell (1 mm), the difference
between these measured widths is a significant portion of the true
pulse width. There have been reports of time-narrowed transient-
stimulated Raman-scattered pulses in this same configuration [21],
and recently Colles has observed stimulated pulses of less than 0.3
psec duration in a Raman oscillator pumped with 530 nm picopulses
[22]. Of course, the light appearing in the camera is the integrated sum of the 15-20 stimulated light pulses which appear in the
train. Observations with a photodiode and a Tektronix 519 revealed
that the 530 nm picopulses in the first part of the mode-locked
train of pulses were more effective in stimulating the Raman pulses
than those in the latter half of the train. This effect probably
arises out of the fact that the initial pulses are of shorter duration and are spectrally narrower than those pulses in the last part

of the train [2]. Time-resolved stimulated Raman scattering was
also observed in two Stokes Raman lines of methanol (ν_1 = 2834 cm^{-1},
ν_2 = 2944 cm^{-1}) with substantially the same results as those obtained
with ethanol.

As we noted above, picopulses can also be obtained in stimulated
fluorescence emission simply by focusing the 530 nm picopulses into
a fluorescent dye. We have used the cross-beam light shutter to observe stimulated emission in rhodamine 6G and acridine red dissolved
in methanol. A typical microdensitometer trace of stimulated fluorescence emission in rhodamine 6G is shown in Fig. 18(c). The apparent half-width is 11 psec, so the true duration is about 6 psec. No
time delay relative to the 530 nm pumping pulse was observed within
the experimental error (±3 psec). All the pulses in the train were
equally effective in stimulating the emission, indicating that there
was little or no buildup of triplet states to form a "bottleneck" for
stimulated fluorescence, and no dependence on the spectral and temporal broadening of the 530 nm picopulses used to pump the dye.

The wavelength of the stimulated emission in rhodamine is concentration dependent as shown in Fig. 19. This concentration dependence indicates that self-absorption of the stimulated fluorescence
emission is responsible for the wavelength shifts. An interesting
aspect of the wavelength dependence is that although the spontaneous
fluorescence spectrum is severely distorted by self-absorption at
these concentrations in the 5 cm cell, so that the apparent maximum
of fluorescence occurs at 575 nm, the stimulated emission appeared
at shorter wavelengths closer to the true maximum of fluorescence in
a dilute solution, 546 nm. This would indicate that the pumping
pulse nearly depletes the ground state, so that self-absorption
effects are decreased for stimulated fluorescence emission.

The spectral width of the stimulated fluorescence emission in
rhodamine is 2 nm, independent of concentration. This is considerably narrower than the spectral width of spontaneous fluorescence,
which is 35 nm (full width at half maximum). It is also somewhat
narrower than the spectral width of the pumping 530 nm picopulses,
namely, 5 nm. This narrow spectral width is not brought about by

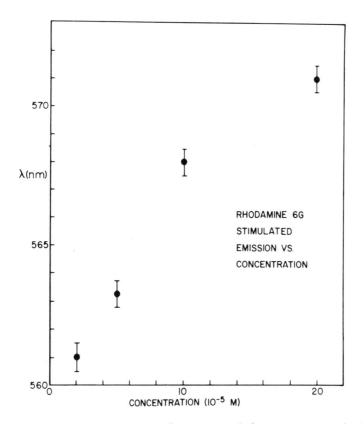

FIG. 19. The wavelength of stimulated fluorescence emission
depends upon the concentration as shown. The most efficient conversion takes place at 5 x 10^{-5} M. At 1 x 10^{-5} M the stimulated Raman
scattering competes with the dye emission. At 5 x 10^{-5} M stimulated
emission is not seen. The concentration dependence of the stimulated
emission is readily explained by self-absorption effects.

any tuning element, but apparently arises out of the gain profile
of the excited state.

Stimulated fluorescence emission was also observed in acridine
red. The apparent time duration of these pulses in the cross-beam
shutter is 14 psec, so the true time duration is about 9 psec. The
emission was centered at 578 nm at a concentration of 6 x 10^{-5} M.
This is also on the short wavelength side of the distorted spontan-

eous fluorescence maximum, 588 nm, and closer to the true fluorescence peak in a dilute solution, 565 nm. The spectral width of the
stimulated fluorescence emission is 9 nm, which is narrower than the
width of the spontaneous fluorescence, 40 nm, and wider than the
spectral width of the pumping pulses. The broader spectral width
(relative to that seen in rhodamine 6G) might be attributed to a
faster vibrational relaxation (relative to the population inversion
time and lasing action) or an inherently broader gain profile (because of Franck-Condon overlap factors).

The limitations of using film in the camera for the cross-beam
shutter are obvious. Because of the lack of sensitivity of the
film and the brief "open" time of the shutter (combined with the
low peak transmission), only those phenomena which produce an intensely bright light pulse can be studied. This limitation can be overcome with an array of electronic photodetectors, such as photoFET's
which combine high sensitivity with compactness. Another possibility
is to use an image-intensifier tube in the place of the film.

Despite the limitations of using film, we were able to combine
the cross-beam light shutter with a spectrograph to obtain time- and
spectrum-resolved emission as shown schematically in Fig. 20. The
presentation on the film is light intensity versus time (measured

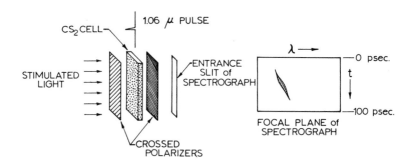

FIG. 20. The time-resolved spectral emission may be determined
by placing a spectrograph behind the cross-beam light gate as
shown. The wavelength dependence is displayed in the direction
perpendicular to the entrance slit as usual, and the time dependence
is displayed in the direction parallel to the surface slit.

along the length of the entrance slit) and wavelength (perpendicular
to the entrance slit in the usual way). The time-resolved spectral
emission of the 530 nm picopulses is shown in Fig. 21(a). Note that
this spectrum represents the integrated light intensity of the 25-30
pulses in the mode-locked train, although the higher intensity
pulses are weighted more heavily because they would see a higher
peak transmission in the shutter. Note that the wavelength shifts
from longer to shorter wavelengths from the beginning to the end of
the pulse. This is consistent with the early measurements of Treacy
of the frequency sweep of the 1.06 μ picopulses [3]. It is also
consistent with what one might expect when a short duration pulse

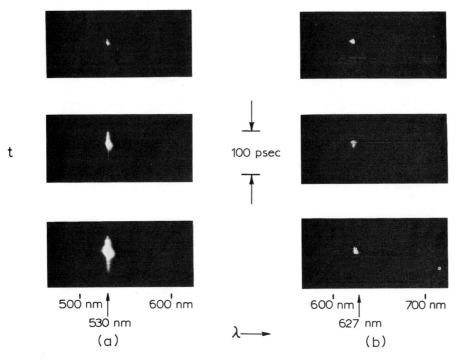

FIG. 21. Three exposures of the time-resolved spectra of (a)
530 nm picopulses and (b) stimulated Raman scattering in ethanol.
The wavelength sweep of the 530 nm picopulses is clearly indicated.
The results of the stimulated Raman scattering are less consistent,
although no sweep is apparent.

(which is made of a group of wavelengths) travels in a medium with normal dispersion of the index of refraction.

The time-resolved spectral emission of transient-stimulated Raman scattering was also observed, and is shown in Fig. 21(b). The reproducibility of this spectrum was exceedingly poor, but there did not appear to be any frequency sweep as in the 530 nm picopulses which were used to pump the Raman pulses. The reasons for this type of behavior are not understood. We also attempted to observe the time-resolved spectra of stimulated fluorescence in rhodamine 6G, but were unable to obtain spectra, probably because of insufficient light intensity.

Topp [36] has demonstrated a nearly collinear geometry for obtaining time-resolved emission spectra which has the advantage of increased transmission through the light gate, without a sacrifice of time resolution. The technique is shown schematically in Fig. 22. The weak light beam is passed through a stepped delay (which is a stack of microscope slides) and travels nearly parallel to the 1.06 μ gate picopulses. This same geometry was adapted by Auston [4] to measure frequency sweeps in the 1.06 μ pulses.

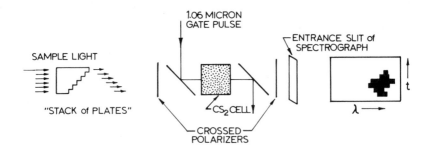

FIG. 22. The echelon or "stack of plates" technique for determining time-resolved spectra. This technique allows the spectral dependence to be determined in the same way as the cross-beam technique but offers higher peak transmission.

Bradley, Liddy, and Sleat [43] recently demonstrated an electronic streak-image camera which is capable of picosecond resolution. A fast photodiode triggers a high-voltage pulse which sweeps the electron image across the screen. This camera offers a linear response independent of the intensity and duration of the 1.06 μ "gate" picopulse used in the traveling-wave Kerr cell, and has more flexibility of application.

VI. CONCLUSION

If a photochemist were to design the ideal instrument for his laboratory, he might include a laser light source of picosecond pulses continuously tunable over the ultraviolet, visible and near infrared spectrum and a streak-image photodetector capable of detecting single photons with picosecond time resolution. With such an apparatus both temporal and spectral information would be gathered simultaneously. Of course the ideal instrument of this sort is not yet available, but with recent developments in the fields of continuous mode-locked dye lasers [5,6] streak-image cameras [43], and photon-counting image detectors [44], one can see that all the elements of such an instrument are or will shortly become reality.

At present, mode-locked neodymium:glass and ruby lasers are temperamental light sources with poor reproducibility and rather ill-defined characteristics. Certainly the experiments outlined here are simple-minded, crude applications of picosecond pulses. As more is learned about producing consistent picopulses and better mode-locking dyes become available, one can expect that more sophisticated picopulse experiments will be performed. It seems inevitable that neodymium:glass and ruby lasers will eventually be displaced by the more versatile dye lasers. It is interesting that flashlamps developed for flash photolysis were used in the first ruby laser, that dyes synthesized by photochemists are used for mode-locking, and that now fluorescent dyes hold the most promise for lasing media.

It would seem that photochemistry has contributed as much to laser knowledge and technology as lasers will ultimately contribute to photochemistry.

ACKNOWLEDGMENTS

We wish to thank D. H. Auston and M. A. Duguay of Bell Telephone Laboratories, M. J. Colles of Heriot-Watt University, Scotland, J. Mitschele of Pennsylvania State University, A. Saxman of Scandia, and K. B. Eisenthal of IBM Research Laboratories for many helpful discussions of picosecond laser techniques. Acknowledgment is made to the donors of the Petroleum Research Fund, administered by the American Chemical Society, to the Research Corporation, and to the National Science Foundation for support of this research and the preparation of this chapter.

NOTE ADDED IN PROOF

Recent experimental studies of Nd:glass mode-locked lasers have been reported by D. von der Linde, IEEE J. Quantum Electron., QE-8, 328 (1972); P. G. Kryukov and V. S. Letoknov, ibid., QE-8, 766 (1972); R. C. Eckardt, ibid., QE-10, 48 (1974). Mode-locked dye lasers have been described by E. G. Arthurs, D. J. Bradley, and A. G. Roddie, Appl. Phys. Letters, 20, 125 (1972); E. P. Ippen, C. V. Shank, and A. Dienes, ibid., 21, 348 (1972); C. V. Shank and E. P. Ippen, ibid. (1974) (in press). Picosecond molecular relaxation in liquids has been studied by A. Laubereau, D. von der Linde, and W. Kaiser, Phys. Rev. Letters, 28, 1162 (1972); R. R. Alfano and S. L. Shapiro, ibid., 29, 1655 (1972); D. Ricard, W. H. Lowdermilk, and J. Ducuing, Chem. Phys. Letters, 16, 617 (1972); G. Mourou, B. Drouin, M. Bergeron, and M. Denariez-Roberge, IEEE J. Quantum Electron., QE-9, 745 (1973); A. Laubereau, L. Kirschner, and W. Kaiser, Opt. Commun., 9, 182 (1973); M. M. Malley and G. Mourou, Opt. Commun. (1974) (in press).

REFERENCES

1. M. A. Duguay, J. W. Hansen, and S. L. Shapiro, IEEE J. Quantum Electron., QE-6, 725 (1970); A. J. DeMaria, W. H. Glenn, and M. E. Mack, Phys. Today, 24, No. 7, 19 (1971); D. J. Bradley, Proceedings of the Technical Programme, Electro-Optics '71 International Conference (1971), p. 1.

2. R. C. Eckardt, C. H. Lee, and J. N. Bradford, Appl. Phys. Letters, 19, 420 (1971).

3. E. B. Treacy, Appl. Phys. Letters, 17, 14 (1970).

4. D. H. Auston, Opt. Commun., 3, 272 (1971).

5. A. Dienes, E. P. Ippen, and C. V. Shank, Appl. Phys. Letters, 19, 258 (1971).

6. D. J. Kuizenga, Appl. Phys. Letters, 19, 260 (1971).

7. E. G. Arthurs, D. J. Bradley, and A. G. Roddie, Appl. Phys. Letters, 19, 480 (1971).

8. W. R. Ware, in Creation and Detection of the Excited State (A. A. Lamola, ed.), Vol. 1, Part A, Marcel Dekker, New York, 1971, pp. 213-300.

9. J. A. Giordmaine, P. M. Rentzepis, S. L. Shapiro, and K. W. Wecht, Appl. Phys. Letters, 11, 216 (1967).

10. G. D. Boyd and H. Kogelnik, Bell Syst. Tech. J., 41, 1347 (1962).

11. R. C. Pastor, H. Kimura, and B. H. Soffer, J. Appl. Phys., 42, 3844 (1971).

12. R. A. Fisher, Lawrence Radiation Laboratory, Report #UCRL-19147 (Jan. 1970).

13. K. B. Eisenthal (private communication).

14. D. C. Burnham, Appl. Opt., 9, 1727 (1970).

15. D. C. Burnham, Appl. Phys. Letters, 17, 45 (1970).

16. D. H. Auston, Appl. Phys. Letters, 18, 249 (1971); D. J. Bradley, B. Liddy, and W. E. Sleat, Opt. Commun., 2, 391 (1971); L. Dahlstrom, Opt. Commun., 3, 399 (1971); D. von der Linde and A. Laubereau, Opt. Commun., 3, 279 (1971).

17. M. Goeppert-Mayer, Ann. Phys., 9, 273 (1931).

18. W. Kaiser and C. G. B. Garrett, Phys. Rev. Letters, 7, 229 (1961).

19. M. A. Duguay and J. W. Hansen, IEEE J. Quantum Electron., QE-7, 37 (1971).

20. M. J. Colles, Opt. Commun., 1, 169 (1969).

21. R. L. Carman, F. Shimizu, C. S. Wang, and N. Bloembergen, Phys. Rev., A2, 60 (1970).

22. M. J. Colles, Appl. Phys. Letters, 19, 23 (1971).

23. M. E. Mack, R. L. Carman, J. Reintjes, and N. Bloembergen, Appl. Phys. Letters, 16, 209 (1970).

24. M. M. Malley and P. M. Rentzepis, Chem. Phys. Letters, 7, 57 (1970).

25. W. H. Glenn, M. J. Brienza, and A. J. DeMaria, Appl. Phys. Letters, 12, 54 (1968).

26. M. E. Mack, Appl. Phys. Letters, 15, 166 (1969).

27. R. R. Alfano and S. L. Shapiro, Phys. Rev. Letters, 24, 1217 (1970).

28. R. R. Alfano and S. L. Shapiro, Phys. Rev. Letters, 24, 592 (1970).

29. R. R. Alfano and S. L. Shapiro, Chem. Phys. Letters, 8, 631 (1971).

30. D. W. Faries, P. L. Richards, J. W. Shelton, V. R. Shen, and K. H. Yang, Intern. Quantum Electron. Conf., Kyoto, Japan, 1970, Paper 8-6; T. Yajima and N. Takeuchi, ibid., Paper 8-5.

31. R. I. Scarlet, J. F. Figueira, and H. Mahr, Appl. Phys. Letters, 13, 71 (1968).

32. J. W. Shelton and J. A. Armstrong, IEEE J. Quantum Electron., QE-3, 696 (1967).

33. K. B. Eisenthal and K. H. Drexhage, J. Chem. Phys., 51, 5720 (1969); T. J. Chuang and K. B. Eisenthal, Chem. Phys. Letters, 11, 368 (1971).

34. M. M. Malley and P. M. Rentzepis, Chem. Phys. Letters, 3, 534 (1969).

35. P. B. Mauer, Opt. Spectra, 4th Quarter, 61 (1967).

36. M. R. Topp, R. M. Rentzepis, and R. P. Jones, Chem. Phys. Letters, 9, 1 (1971).

37. P. M. Rentzepis, Chem. Phys. Letters, 2, 117 (1968).

38. E. Drent, G. M. Van der Deijl, and P. J. Zandstra, Chem. Phys. Letters, 2, 526 (1968).

39. R. A. McLaren and B. P. Stoicheff, Appl. Phys. Letters, 16, 140 (1970).

40. G. Mayer and F. Gires, Compt. Rend., 258, 2039 (1964).

41. F. Shimizu and B. P. Stoicheff, IEEE J. Quantum Electron., QE-5, 544 (1969).

42. M. A. Duguay and J. W. Hansen, Appl. Phys. Letters, 15, 192 (1969); Opt. Commun., 1, 254 (1969).

43. D. J. Bradley, B. Liddy, and W. E. Sleat, Opt. Commun., 2, 391 (1971).

44. E. W. Dennison, Science, 174, No. 4006, 240 (1971).

CHAPTER 4

DYE LASERS

Andrew Dienes and Charles V. Shank

Bell Laboratories
Holmdel, New Jersey

Anthony M. Trozzolo

Bell Laboratories
Murray Hill, New Jersey

I. INTRODUCTION

Ever since the invention of the first laser, the search for new laser materials has continued. While the possibility of obtaining amplification in solutions of organic molecules had been noted rather early [1] it was not until 1966 that laser action in organic dyes was achieved. Sorokin and Lankard [2] reported laser emission from a solution of chloraluminum phthalocyanine which was excited by intense pulses from a giant-pulse ruby laser. Independently, Schäfer, Schmidt, and Volze [3] obtained laser action in the near infrared from a number of cyanine dyes. Soon a number of other dyes were made to lase; Schäfer, Schmidt, and Marth were the first to report dye laser operation in the visible region [4]. Since then, there has been a literal explosion in the development of dye lasers as well as in the understanding of the photophysics of dye molecules [5-7]. The present chapter surveys the achievements in both of these areas and describes the current state of the art.

The intense interest in dye laser research has been motivated largely by the potential for developing sources of tunable coherent radiation. As light sources, the fixed-frequency lasers which were developed in the last decade provided high intensity, extreme monochromaticity, and superb collimation and coherence. These properties have been applied to a variety of research in Raman scattering,

Rayleigh scattering and nonlinear optical interactions. However,
in many spectroscopic or photochemical applications, the fixed-
frequency lasers have one serious limitation: the laser emission
occurs only at certain discrete wavelengths that are characteristic
of the lasing material and therefore are not at the disposal of the
chemist.

The tunability of the dye laser is a consequence of the fact
that dyes have a broad continuous fluorescence spectrum rather than
one or a series of narrow discrete fluorescent emission lines
characteristic of other lasing materials. In the first dye laser
experiments, no attempt was made to control the wavelength and
bandwidth of the laser oscillations. Laser action occurred at the
peak of the fluorescence curve with a rather wide ($\sim$50 $\overset{\circ}{\text{A}}$) bandwidth.
By replacing one of the resonator mirrors with a diffraction grating,
Soffer and McFarland [8] succeeded in tuning the wavelength of a dye
laser as well as narrowing the oscillation bandwith to 0.6 $\overset{\circ}{\text{A}}$. Most
important, the power output in this narrow-band, tunable operation
was substantially the same as in the wide-band operational mode.

Since organic dye solutions have a broad absorption spectrum,
pumping of dye lasers by incoherent sources, such as high-intensity
flashlamps, is also feasible. The first flashlamp-pumped dye laser
was reported by Sorokin and Lankard [9]. Thereafter, a considerable
number of other dyes were made to lase in flashlamp systems. The
range of wavelengths covered by dye lasers was extended by new dye
laser materials such as organic scintillators. Today the tunable
dye laser provides continuous coverage of the electromagnetic spectrum from the near ultraviolet (3200 $\overset{\circ}{\text{A}}$) to the near infrared
(1.2 μ). Spectral purity down to 1 Mhz bandwidth has been achieved.
For some applications, continuous-wave (cw) operation of the laser
is desirable. Until recently, both flashlamp and laser-pumped dye
lasers operated only on a pulsed basis. The first cw dye laser
operation was achieved by Peterson, Snavely, and Tuccio [10] using
a rhodamine 6G solution and pump light from an argon laser. A
number of other dyes have since been lased on a continuous basis,
and at present, tunable cw dye-laser light can be generated from
5200 to 7000 $\overset{\circ}{\text{A}}$ [11,12].

Another desirable feature of dye lasers is their potential for
the production of ultrashort (picosecond) pulses. The ultimate ob-
tainable pulsewidth is of the order of $\tau \approx (\Delta\nu)^{-1}$, where $\Delta\nu$ is the
bandwidth measured in frequency units. Even if a bandwidth of only
about 30 cm^{-1} is utilized, pulses of the order of 1 psec duration
can be generated. The best and most convenient method for the
generation of ultrashort pulses is to place a suitable saturable
absorber in the laser cavity. This method was first used successfully with a flashlamp-pumped dye laser by Schmidt and Schäfer [13]
and later by Bradley and coworkers [14,15], who have generated
bursts of pulses of approximately 5 psec duration, tunable over
the wavelength range 5840-7040 Å. Very recently the cw rhodamine
6G dye laser also has been mode-locked to produce a continuous
stream of 1.5 psec pulses [16].

In the recent years the technology of dye lasers and the preparation and purification of dyes has progressed to the point that
the dye laser has become a valuable tool for the experimental chemist. A corollary effect in seeking better performing dye lasers
is the increased understanding of the spectroscopy of dye molecules,
since it is the excited-state properties of the dyes which determine not only the laser characteristics but dye stability as well.

This chapter is divided into six parts. In the second part,
the spectroscopy of organic dye molecules is reviewed and the
classes of compounds useful as dye laser materials are described.
In the third part, we describe the basic principles of the laser
and show how organic dye molecules can be made to develop optical
gain. The necessary pump power densities are calculated and expressions are derived for the dye laser gain as a function of the
dye properties. The fourth part is devoted to a description of
practical dye laser devices using a variety of pumping techniques
and laser configurations. In the fifth part we describe a number
of novel and useful applications of dye lasers that have been recently reported in the literature. We conclude with a brief look
at the future of dye laser technology and its applications.

II.   PHOTOPHYSICS OF DYES

A. Properties of Organic Dyes

Dyes, according to the commonly accepted definition, are colored substances having the ability to impart their color to other materials. However, the present usage of the term in a variety of technologies no longer requires that the substance be colored, and materials which are called dyes today can absorb or emit in the ultraviolet and near infrared wavelength ranges as well as in the visible. For example, the so-called "whiteners" which are used on many fabrics absorb in the UV region and emit blue light. Also, infrared-absorbing dyes are essential in the operation of Q-switched neodymium lasers.

Although many thousands of dyes are known, only a relatively few fluoresce in solution and therefore can be considered for use in lasers. Of primary importance are the spectroscopic properties which depend on the chemical structure of the dye as well as on any interaction with the solvent. In the next section, we discuss the spectroscopy of dyes, classes of lasing dyes, and the dye photochemistry which is pertinent to laser technology.

B. Spectroscopy

The optical absorption and emission properties of organic dyes have been studied for many years and have been the subject of numerous review articles [17-19]. As expected, the features of the spectral process (wavelength range, width, band structure, and intensity) vary for different dyes or for particular solvent environments. However, there are certain general characteristics of most fluorescing dyes. These include (see Fig. 1):

1. Both the absorption and emission bands are fairly broad ($\sim$15,000 cm^{-1}).

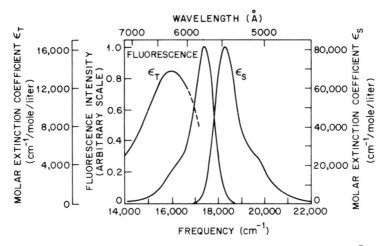

FIG. 1. Absorption and fluorescence spectra of 5 x 10^{-5} M Rhodamine B in methanol.

2. The fluorescence band is generally a mirror image of the absorption band.

3. The fluorescence lifetime is about 10^{-9} sec.

4. The maximum of the fluorescence band occurs at longer wavelengths (lower energy) than the maximum of the principal absorption band. This displacement is called the Stokes shift of fluorescence from the absorption, and if it is small, there may be substantial overlap between the two bands.

5. A triplet-triplet absorption band may overlap the fluorescence band (see below).

In order to explain these processes in terms of the excited states of the dye and thus gain insight into the laser process in organic dyes, it is worthwhile to employ the energy-level diagram shown in Fig. 2. The electronic ground state of the dye molecule is a singlet state S_0, which includes a range of energies which are determined by the quantized vibrational and rotational states of the molecule. The energy difference between neighboring vibrational states (indicated in the figure by heavy horizontal lines) is about 100-3000 cm^{-1} while the energy spacing between rotational levels is

smaller by approximately two orders of magnitude. The rotational
levels therefore form a near continuum of states between the vibrational levels, and it is possible to describe each electronic state
of the molecule in terms of a similar broad continuum. Optical
transitions between these continua then give rise to the characteristic broad absorption and emission bands. In Fig. 2, the first
and second excited singlet states are designated S_1 and S_2.

As the first step in the laser process, the dye molecules are
excited from the lowest levels of the ground singlet state S_0 to
higher vibrational-rotational levels of the S_1 state by the absorbed
light (A → b). The energy of the molecule very quickly decays
($\sim 10^{-11}$ sec) in a nonradiative process to level B, the lowest vibrational level of the S_1 state (b → B). (Strictly speaking, thermal
equilibrium is established quickly).

A molecule in the lowest level of S_1 can return to S_0 by emitting a photon of light (B → a) whose energy is less than that of
the absorbed light. This radiative process (fluorescence) is thus
shifted to longer wavelengths from the absorption.

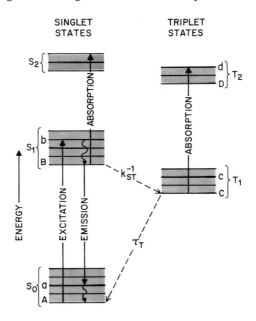

FIG. 2.   Energy levels of a dye molecule.

Since fluorescence is a spin-allowed $(S_1 \to S_0)$ process, a typical fluorescence lifetime for organic dyes is about 10^{-9} sec, although some variation in the radiative lifetime for a particular dye can be effected by changing solvent, concentration, or temperature of the solution. In contrast, the corresponding decay time for solid-state and inorganic laser materials is many orders of magnitude slower ($\sim 10^{-3}$ sec). The ratio of the number of emitted photons to the number of absorbed pump photons is called the fluorescence quantum efficiency and can be as high as unity for some dye solutions or essentially zero for other materials. It is clear that a high fluorescence quantum efficiency is desirable in a dye laser system.

In addition to fluorescence, a molecule in S_1 can undergo three other types of transitions which therefore compete with the radiative process. Transitions between S_1 and other excited singlet states $(S_1 \to S_2)$ can possibly occur at the same wavelength as the fluorescence. Also, nonradiative transitions between states of the same multiplicity (internal conversion) can occur $(S_1 \leadsto S_0)$ and thus reduce the fluorescence quantum efficiency. Molecules in the excited singlet S_1 state may undergo a nonradiative process with change in multiplicity to a lower lying triplet state T_1 instead of decaying to the ground state. This process (intersystem crossing) proceeds at a rate governed by an intersystem crossing rate constant k_{ST}.

Intersystem crossing reduces the population of S_1 available for radiative transition to S_0 and so reduces the fluorescence quantum efficiency. The lifetime for decay of the triplet state $(T_1 \to S_0)$ is generally much longer than the fluorescence lifetime, since the triplet-singlet transition is spin-forbidden. Owing to its relatively long lifetime, the triplet state acts as a trap for the excited molecules, which then accumulate in the T_1 level.

The state T_1 is the lowest-lying state of a manifold of excited triplet states, of which T_2 is the next highest. Triplet-triplet absorptions $(T_1 \to T_n)$ are spin-allowed and usually have relatively high extinction coefficients. In a great number of fluorescent dyes, a triplet-triplet absorption band overlaps with the $S_1 \to S_0$ spectrum

(see Fig. 1). It will be shown later that the accumulation of
molecules in the triplet T_1 state can produce a large optical loss,
detrimental to laser action.

It is to be noted also that the triplet state T_1 of many molecules is a chemically reactive species. As a result, a permanent
destruction of the dye may occur. The consequence of this photochemical behavior is that, although the quantum yield for the photochemical reaction may be very low ($\sim 10^{-6}$ at 77°K in a glassy matrix
[20,21] the light intensities required for laser action are sufficiently high (see Sec. III) so that a rigid dye solution (such as
in a plastic matrix) cannot be used as the laser-active material
for any prolonged period.

C. Factors Influencing Fluorescence

The most important factor influencing laser operation of an
organic dye is its fluorescence. In this section, we shall discuss
the various factors which can influence the fluorescence process.
Included are the effects of (a) substituents, (b) geometry and structure, (c) solvent, (d) pH, and (e) quenching.

1. Substituent Effects

Certain functional groups on an aromatic ring can exert defi-
nite and predictable effects on the energy and intensity of the
fluorescence of the molecule. These effects are summarized in Table
1 [22], and a few of them are discussed below.

a. Alkyl groups. In general, addition of alkyl groups of an
aromatic nucleus has little influence on the energy or intensity
of the fluorescence. However, in a number of aromatic systems, sub-
stitution of methyl or ethyl groups causes a substantial increase
in intersystem crossing and thus cuts down the fluorescence efficiency [23]. Presumably steric factors are involved.

TABLE 1

Substituent Effects on Fluorescence of Aromatic Molecules

Substituent	Emission peak shift	Effect on intensity
Alkyl	None	Very slight increase or decrease
$-OH$, $-OCH_3$, $-OC_2H_5$	Red shift	Increase
$-CO_2H$	Red shift	Large decrease
$-NH_2$, $-NHR$, $-NH_2$	Red shift	Increase
$-NO_2$, $-NO$	--	Total quenching
$-CN$	None	Increase
$-F$, $-Cl$, $-Br$, $-I$	Red shift	Decrease
$-SO_3H$	None	None

    b. <u>Halogens.</u> In a study of halogen-substituted naphthalenes, McClure [24] found that the phosphorescence-to-fluorescence ratio ϕ_p/ϕ_F increased as one goes through the series, F, Cl, Br, and I. This effect has been termed the intramolecular heavy-atom effect. In a study of the mechanism of the heavy-atom effect, it has been suggested that the halogen substituent introduces a new spin-orbit coupling mechanism as well as enhancing that of the aromatic molecules [25]. The heavy atom need not be a substituent of the emitting molecule in order to exert this influence; a heavy-atom-containing solvent will also quench the fluorescence of the solute (see Sec. II. C.5).

    c. <u>Nitro groups.</u> In general, nitro aromatic compounds do not fluoresce. It is thought that heavy-atom effects and predissociation are probably both involved in the quenching of fluorescence by the nitro group.

    d. <u>Hydroxyl and amino groups.</u> These groups are of great interest because of the unusual acid-base properties of the excited

states of aromatic compounds containing these groups. These effects
are therefore described in a separate section (II. C.4).

2. Effect of Molecular Geometry and Structure

Many intensely fluorescent organic molecules possess a highly
rigid, planar structure (see the structure diagrams). For example,
fluorescein and pyronin Y show intense fluorescence in solution,
whereas phenolphthalein and Michler's Hydrol Blue, which are struc-
turally similar, respectively, do not fluoresce.

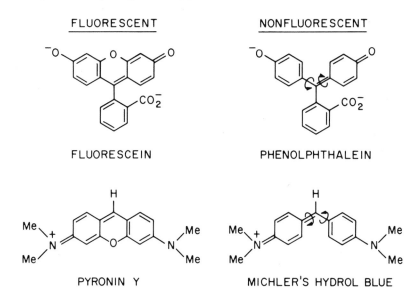

FLUORESCENT	NONFLUORESCENT
FLUORESCEIN	PHENOLPHTHALEIN
PYRONIN Y	MICHLER'S HYDROL BLUE

The only structural difference between the two pairs of dyes is an
oxygen bridge which is present in fluorescein and pyronin Y, but
which is absent in phenolphthalein and Michler's Hydrol Blue. In
the latter two compounds, the rotation and vibration of the aromatic rings relative to each other can occur quite easily, so that
electronic excitation energy can be dissipated readily in a

nonradiative process. Vibrational or rotational dissipation of
excitation energy is considerably more difficult in the rigid structures of fluorescein and pyronin Y. It is therefore possible to say
that, as a general rule, the fluorescence efficiency increases
greatly with the rigidity of the molecular structure.

Steric crowding also facilitates internal conversion and thus
leads to decreased fluorescence [26]. As has been pointed out in
the previous discussion on substituent effects, alkyl groups have
relatively little effect on the fluorescence of aromatic compounds.
However, if the alkyl groups are bulky and can interfere with each
other, it is possible to observe marked decreases in fluorescence
intensity, and examples of such effects are quite common in cyanine
dyes [26].

If the steric hindrance destroys the coplanarity of the molecule,
fluorescence efficiency is decreased sharply, and may in fact be
completely absent. This behavior is particularly apparent for a
number of compounds which exhibit geometrical isomerism, where often
the cis-isomers are nonfluorescent, while the trans-isomer shows intense emission. An example is the pair of isomers of stilbene where
the cis-isomer presumably is nonplanar because of steric interference
of the ortho-hydrogen atoms of the two phenyl rings [27,28].

3. Solvent Effects

The effect of the solvent on fluorescence has not received as
much attention as ultraviolet absorption solvent effects or the effects of pH on absorption and fluorescence spectra (see Sec. II.C.4);
nevertheless, several groups of dyes are known in which a change of
solvent brings about remarkable changes in the fluorescence spectrum.
In some cases, these changes are accompanied by comparable changes in
the fluorescence excitation spectrum (and therefore also in the UV
absorption spectrum). In other cases, the absorption spectrum remains relatively unchanged. Displacements in both the absorption
and fluorescence spectra imply an interaction of the solvent in both

the ground state and the excited state of the molecule. On the
other hand, when the fluorescence emission spectrum alone is changed,
the implication is of an interaction of the solvent with the excited
state of the molecule, but not with the ground state. These effects
have been reviewed [29,30] and are discussed only briefly here.

To understand why the influence of solvent on the fluorescence
spectrum is different from that observed on the absorption spectrum,
one must invoke the Franck-Condon principle, which states that electronic transitions occur much more rapidly than motions of atomic
nuclei. Since the electronic charge distributions in excited states
of dyes are usually quite different from those of the ground-state
molecule, the "equilibrium" or "relaxed" arrangement of solvent molecules about an excited solute molecule may be very different from
the solvent arrangement about the ground-state solute molecule.
Therefore, after electronic excitation has taken place, reorientation of the solvent must occur. The excited state in which solvent
reorientation has not yet occurred is called a Franck-Condon excited
state, and the relaxed state (after solvent reorganization has taken
place) is described as an equilibrium excited state. It is from
this "relaxed" state that fluorescence occurs at room temperature.
By analogy there is a Franck-Condon ground state which persists momentarily until the solvent molecules reorganize to the equilibrium
arrangement for the ground state (see Fig. 3). Since the solvent
reorientation process is a direct result of the polarization of the
excited state of the solute, it is not surprising that solvent shifts
in the fluorescence spectrum have been related to the dielectric constant of the solvent. Bayliss and McRae [31] termed these energy
changes "polarization shifts" and have ascribed them to polarization
of solvent molecules induced by the excited-state dipole of the solute. This area has been reviewed by Lippert [30], and studies of
the electrostatic shifts of fluorescence maxima clearly show that
many organic dyes become considerably more polar when electronically
excited. An example of the change of fluorescence maxima due to a
decrease of the Franck-Condon shift is found in the behavior of

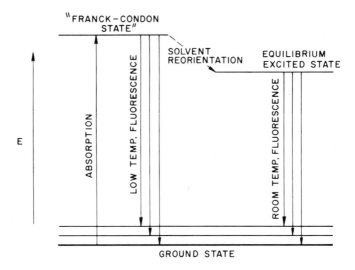

FIG. 3.   Effect of solvent reorientation on a fluorescence band.

Rhodamine 6G in hexafluoroisopropyl alcohol solvent, where a blue
shift in the fluorescence spectrum is observed.   An application of
this effect in cw dye laser technology is discussed in Section IV.C.
     Another very specific solvent-solute interaction is hydrogen
bonding.   An extensive study of hydrogen-bonding effects on fluorescence has been carried out by Mataga, who found that hydrogen bonding can cause either red or blue fluorescence shifts [32].   For example, in a series of naphthylamines in alcohol solvents, it was
possible to show that bonding of an amine hydrogen to a solvent oxygen atom produced a red-shifted fluorescence spectrum, while bonding
of a solvent hydrogen atom to the amine nitrogen atom resulted in a
blue-shifted spectrum.   However, it is difficult to predict the type
of hydrogen bonding that will predominate in the fluorescence spectrum, since the hydrogen-bonding ability of the ground-state solute
molecule may be markedly different from that of the excited state.
It also has been reported that hydrogen bonding may have a pronounced
effect on the fluorescence efficiency of the dye [32].

4. Effect of pH

Aromatic compounds containing amine or hydroxyl substituents
may behave as weak acids or bases. It is therefore expected that
in proton solvents, partial dissociation or protonation will occur
in the ground-states species. This will be reflected in a corresponding change in the absorption and emission spectra (see Sec. V.D).

As mentioned briefly in Sec. II.C.1, the effects of pH on the
fluorescence spectrum of an organic dye often are quite different
from those on the absorption spectrum, particularly if certain functional groups such as the hydroxyl or amino groups are present. In
1931 Weber [33] noted that fluorescence of 1-naphthylamine-4-sulfo-
nate [structure (I) in the accompanying figure] changed color as
the pH of the solution was altered, while no corresponding change

(I)

occurred in the absorption spectrum. This phenomenon was first explained by Förster [34], who found a similar effect on the fluorescence of 3-hydroxypyrene-5,8,10-trisulfonate [structure (II)].

(II)

The pK_a of this compound is approximately 7.3, but Förster noted
that the compound showed a fluorescence characteristic of the pheno-
late anion in solutions which were too acidic for this species to
exist in the ground state. He therefore suggested that phenols must
be stronger acids in their excited states than in their ground states
[34]. This hypothesis has been verified by Weller [35] in a series

of elegant flash-photolysis experiments. Similar changes of acidity
in the excited state have been found for a wide variety of ionizable
organic compounds. A contrasting behavior is found in nitrogen
heterocycles such as acridine [structure (III)], which becomes a

(III)

stronger base ($pK_a^* = 10.6$) in the excited state than in the ground
state ($pK_a = 5.5$): Weller [36] and Vander Doncket [37] have written
excellent reviews summarizing this area and the techniques for determining the excited-state pK_a of a fluorescent system. Some typical
excited-state acid-base properties are given in Table 2 [22,38].

The fact that such acidity or basicity differences can be observed by fluorescence measurements indicates that the proton transfer to or from the solvent must be more rapid than the fluorescence
decay time of the "neutral" molecule ($\sim 10^{-8}$-10^{-10} sec). In some
cases, the proton transfer reaction is so rapid that a state of
equilibrium is attained in the very short interval between the absorption and emission of light. In other cases, the proton transfer

TABLE 2

Acid-Base Properties of Excited States

	pK_a (S_0)	pK_a (S_1)	pK_a (T_1)
Phenol	9.97	5.7	--
p-Cresol	10.27	4.3	--
1-Naphthol	9.23	2.5	--
2-Naphthol	9.46	3.0	8.1
1,2-Naphthalenediol	8.11	2.4	--
3-Hydroxypyrene-5,8,10-trisulfonate	7.30	1.0	--
Acridine	5.5	10.6	5.6
2-Naphthylamine	4.1	-2	3.3
2-Aminoanthracene	3.4	-4.4	3.3

is not sufficiently fast to allow the reaching of equilibrium before
fluorescence occurs. An important application of this behavior in
laser technology is found in the extension of the tuning range of a
dye laser by utilizing excited-state proton transfer reactions (see
Sec. V.D).

5. Quenching

Quenching of fluorescence is defined as any process that results in a decrease in the true fluorescence efficiency of a molecule, i.e., quenching processes divert the absorbed energy into
channels other than fluorescence of the molecule.

The excited-state proton transfer reactions represent one type
of external quenching since, although emission still occurs, it is
not the fluorescence of the absorbing molecule. A similar type of
reaction is association or "excimer" formation in which a change in
the fluorescence spectrum of a given compound occurs with increasing
concentration. Since no corresponding change is observed in the absorption spectrum, the new fluorescence component has been ascribed
to an excited dimer formed only after the absorption of light by the
monomer. This type of fluorescence has been reviewed recently by a
number of authors [39,40].

Excited-state complexes ("exciplexes") of the above type may be
formed by dissimilar molecules also, so that, for example, anthracene fluorescence is quenched in the presence of increasing concentration of diethylaniline with the concomitant occurrence of a new
fluorescence band at about 5000 cm^{-1} to the red of the anthracene
fluorescence. This new fluorescence has been ascribed to electron
transfer in the excited state giving rise to an excited-state donor-
acceptor complex $(D^+A^-)^*$ which emits at longer wavelengths [41].

However, excited-state reactions do not always lead to a new
emitting species. For example, if the electron transfer is to or
from an anion such as I^-, SCN^-, Br^-, or Cl^-, then only quenching

occurs. Solutions of Rhodamine 6G with any of these anions show a
reduced fluorescence quantum efficiency in nonpolar solvents in which
the dye salt is not dissociated. Avoidance of this quenching process
can be achieved by the use of anions such as ClO_4^- or BF_4^-, since
these anions do not undergo electron transfer easily. Drexhage [42]
found that Rhodamine 6G perchlorate showed a fluorescence quantum
yield of about 0.95 which was practically independent of solvent.

In addition to quenching by excited-state reactions, another
very important mechanism by which quenching can occur is energy
transfer from the excited state to another molecule:

$$D^* + A \rightarrow D + A^*$$

There are several distinct processes by which singlet excitation
energy of organic molecules can be transferred in liquid solution:

1. Radiative transfer (fluorescence followed by absorption).
This process is an important one in dye laser technology since, as
mentioned in Sec. II.B, overlap between the emission and absorption
spectra of a given dye causes the tuning range of the dye laser to
be cut off at the short wavelength of the fluorescence.

2. Diffusion-controlled nonradiative transfer.

3. Long-range dipole-dipole resonance interaction.

The solvent also plays an important role in quenching processes.
For example, the fluorescence quantum yield of Rhodamine B, Pyronin
Y can be increased considerably through the use of solvents with
strong molecular dipole moments (e.g., methylene chloride, o-dichloro-
benzene, hexafluoroisopropyl alcohol), which provide a micro rigidity
around the dye molecule [42].

Since transfer of electronic energy into molecular vibrations
(particularly of C-H bonds) results in a decreased fluorescence
yield, it is often possible to increase markedly the fluorescence
efficiency of a dye by deuterating the hydroxyl groups of alcohol
solvents [42].

D. Quenching of Triplet States

Earlier, it was mentioned that intersystem crossing ($S_1 \leadsto T_1$;
Fig. 2) constitutes an undesirable process in the operation of a dye
laser since it is a radiationless process which depletes the population of possible emitting molecules as well as increasing the population of a comparatively long-lived state which can undergo triplet-
triplet absorption ($T_1 \rightarrow T_2$) at wavelengths overlapping with the
fluorescence spectrum of the dye. It is therefore important that we
discuss processes which quench triplet states of dyes, since this
provides an avenue for avoidance of their undesirable effects on the
laser process.

Oxygen is a particularly effective quencher of organic triplets
in solution and probably operates by an energy transfer mechanism.

$$^3D^* + \,^3O_2 \rightarrow D_0 + \,^1O_2$$

Other triplet quenchers include a variety of dienes such as pipery-
lene, 1,3-cyclooctadiene, and cyclooctatetraene. Another very good
quencher of triplet molecules in nitric oxide. Little, if any, use
of this compound in dye laser technology has been reported.

E. Laser Dyes

A survey of the published literature on laser dyes [5-8,43-45]
leads to the following classes of dyes which are useful in dye laser
technology: (a) organic scintillators, (b) coumarins, (c) acridines,
(d) xanthenes, (e) cyanines and (f) polymethine dyes. The tuning
range for the dye lasers derived from dyes of these classes are given
in Table 3 and the structural formulas of typical members are given
in Fig. 4.

Although one usually must investigate the absorption and emission spectra of the particular dye-solvent combination which is of
interest, there are some useful guidelines in the literature. For

TABLE 3

Laser Action of Dyes

Class	Spectral range (nm)
Scintillators	340-435
Coumarin dyes	390-560
Pyrylium salts	450-490
Fluorescein dyes	550-600
Rhodamine dyes	550-650
Cyanine dyes	650-850
Cyanine dyes	1100-1200
Phthalocyanine dyes	750-780

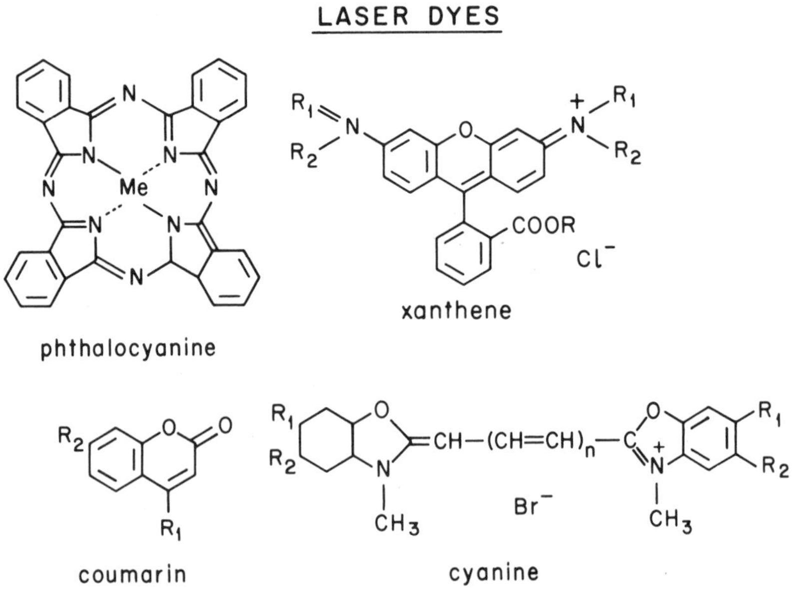

LASER DYES

FIG. 4.   Structural formulas of typical laser dyes.

example, Brooker [46] found that the absorption and emission spectra of a family of polymethine dyes shift to longer wavelengths as the number of -CH=CH- groups in the polymethine chain is increased. The analogous behavior in dye lasers was observed by Miyazoe and Maeda [47]. The xanthene dyes absorb and emit in the visible part of the spectrum. The organic scintillators (oxazoles, anthracenes) and coumarins absorb in the near UV and emit in the violet or blue-green regions of the spectrum.

Chemical or photochemical instability is a very important factor in a number of dyes which otherwise might have potential in laser technology. Miyazoe and Maeda [47] found that chemical instability results when there are more than four -CH=CH- units in a polymethine chain of a cyanine dye. Other factors such as solvent, pH, and concentration also play an important role in determining the stability of the dye.

III. BASIC LASER PRINCIPLES

A. Stimulated Emission, Population Inversion, and Gain

To understand the physical phenomena that make the laser possible let us briefly consider some simplified and idealized systems. First, let us consider a two-level quantum system with energies E_1 and E_2 as shown in Fig. 5(a). Let us assume that there exists an ensemble of N particles in the above system, N_1 being the population of the lower level and N_2 that of the upper. Spontaneous emission of radiation of frequency $\nu = (E_2 - E_1)/h$ takes place from the upper level with probability A_{21}, resulting in an isotropic emission of power $N_2 A_{21} h\nu$ [48,49]. If, in addition, the particles are subjected to a beam of light frequency ν, those in the upper state will emit radiation of the same frequency, polarization, and propagation

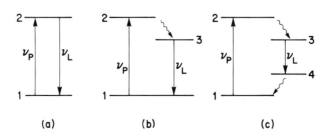

FIG. 5. (a) Two-level laser system; (b) three-level laser system; (c) four-level laser system.

direction. The probability for this <u>stimulated emission</u> is $B_{21}I(\nu)/4\pi$, where $I(\nu)$ is the power density of the stimulating radiation. Those particles in the lower state will absorb radiation with the probability $B_{12}I(\nu)/4\pi$, where $B_{12} = B_{21}$ for nondegenerate levels. The net probability for the absorption of a photon from the stimulating beam is

$$p = \frac{B_{12}I(\nu)}{4\pi}(N_1 - N_2) \tag{1}$$

From this it is very simple to derive the absorption coefficient k, where k is defined by the well-known Beer's law for the propagation of radiation through a medium [48,49]

$$I(z) = I_{in}e^{-kz} \tag{2}$$

Neglecting spontaneous emission, k is found to be

$$k = (h\nu/4\pi)B_{12}(N_1 - N_2) \tag{3}$$

The normal situation in a medium is that $N_1 > N_2$. In that case k is positive and we have resonant absorption. However, if by some means it is possible to invert the population and make $N_2 > N_1$, then we have negative absorption or gain. Such a population-inverted medium with gain at optical wavelengths is a laser amplifier.

The two-level system of Fig. 5(a) is not very suitable for achieving gain. In fact, under steady-state optical excitation the absorption coefficient of such a system is always positive. It is possible, however, to obtain population inversion by optical excitation between two levels of three- and four-level systems, such as are shown in Fig. 5(b) and (c). The ruby laser is an example of a three-level system, while glass lasers and solutions containing rare earth atoms are four-level systems. For each of these cases the upper laser level must be long lived, and the laser linewidth is narrow.

Gain in organic dyes which have very broad absorption and emission linewidths can also be achieved by optical pumping. The mechanism by which an effective population inversion is achieved is somewhat different and is described in the following section.

B. Dye Lasers

In organic dye solutions, laser action takes place between the ground singlet state and the first excited singlet state electronic levels, both of which are substantially broadened by a vibrational-rotational continuum. We now show that it is possible to achieve amplification in these systems by optical pumping, even when much less than 50% of the molecules are raised to the excited singlet state.

Consider the two lowest singlet levels of the energy-level diagram in Fig. 2. Intense pumping radiation excites N_1 molecules into the excited singlet state. Thermal redistribution among the continuum of sublevels takes place within a very short time (of the order of 10^{-11}-10^{12} sec). The gain coefficient at the frequency ν then is

$$\alpha(\nu) = -k(\nu) = (h\nu/4\pi)[N_1 B_{10}(\nu) - N_0 B_{01}(\nu)] \qquad (4)$$

For the organic dye solution $B_{10}(\nu)$ does not equal $B_{01}(\nu)$ because
of the Boltzmann distribution of the particles within the continuum
and because of the Stokes shift of the emission. In some range of
frequencies B_{10} can be substantially larger than B_{01} and amplification can be achieved even when only a few per cent of the molecules
are in the excited state. It is convenient to express Eq. (4) in
terms of the absorption and emission cross sections $\sigma_a(\lambda)$ and $\sigma_e(\lambda)$
of the dye molecules. Thus

$$\alpha(\lambda) = N_1 \sigma_e(\lambda) - N_0 \sigma_a(\lambda) \qquad (5)$$

Fig. 6 shows these cross-sections as a function of wavelength for
the laser dye Rhodamine 6G. The absorption cross section is a readily measured quantity, while the emission cross section is derived
from the spontaneous emission (fluorescence) spectrum by the relationship [50,51]

$$\sigma_e(\lambda) = \frac{\lambda^4 E(\lambda)}{8\pi \tau c n^2} \qquad (6)$$

Here n is the index of refraction, τ is the fluorescence decay
rate, and $E(\lambda)$ is the true emission line-shape normalized such that

$$\int_0^\infty E(\lambda)\, d\lambda = \phi \qquad (7)$$

where ϕ is the quantum efficiency of fluorescence. From Eq. (5) and
Fig. 6 it is evident that at the long-wavelength side of the dye
emission it is possible to obtain gain even when $N_1 < N_0$.
    The above description of a dye laser is a simplified and incom-
plete one. In an actual dye laser system the additional energy
levels always tend to limit and can actually prevent laser action.
The most important detrimental effect is that due to the triplet
manifold. Intersystem crossing to the triplet manifold reduces the
quantum efficiency since it is a nonradiative process. It therefore
goes without saying that only dyes with reasonably high quantum efficiency are suitable laser substances. An even more serious effect

of intersystem crossing is due to the fact that for a great number
of fluorescent dyes the triplet-triplet absorption band overlaps
the emission band and thus acts as an additional loss mechanism.
Triplet-triplet absorption usually has a relatively large cross section. Owing to its long lifetime the triplet state acts as a trap
for the excited molecules, and the accumulation of the molecules in
the T_1 state produces a large optical loss at the wavelength where
laser emission would take place. It will be shown later that this
can be a serious limitation even for dyes with high quantum yield.

Additional losses at the laser wavelengths may be present because of excited-state absorption (e.g., $S_1 \rightarrow S_2$). There exist
a number of organic compounds with fluorescence quantum yield close
to unity for which attempts to achieve laser action have been unsuc-
cessful. (Rubrene is one example.) This is thought to be caused

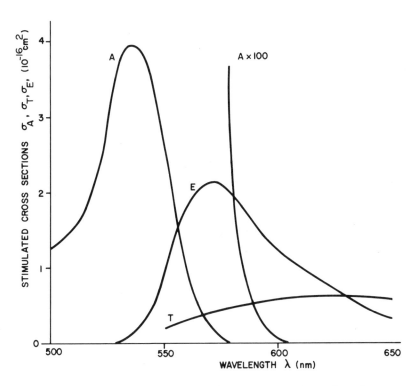

FIG. 6. Emission and absorption cross sections for the laser
dye Rhodamine 6G.

by strong excited-singlet absorption overlapping the emission bands. Taking into account the above factors the gain of a dye laser amplifier is

$$\alpha = N_1\sigma_e - N_0\sigma_a - N_T\sigma_T - N_1\sigma_a^* \qquad (8)$$

where N_T is the population of the lowest triplet state, σ_T is the triplet absorption cross section, and σ_a^* is the excited singlet absorption cross section (e.g., $S_1 \rightarrow S_2$) in Fig. 2. Other quantities are as defined previously.

So far we have discussed the concept of optical gain in organic dyes. A medium with gain at optical wavelengths is a laser amplifier. Under some pumping conditions extremely high gains ($\alpha L \approx 10$ or higher) can be achieved in organic dyes. In such cases strongly amplified spontaneous emission can result in intense laser output. The emission intensity at one end of such a high-gain laser amplifier is given by

$$I(\lambda) = C\ E(\lambda)\ \exp^{[\alpha(\lambda)L]} \qquad (9)$$

where C is a constant depending on geometry. Since α is wavelength dependent, it is obvious that because of the exponential gain factor, the output line-shape can be strongly narrowed if α is high.

A more usual laser source is a laser oscillator, achieved by providing an optical resonant cavity around the gain medium. The emitted radiation is contained within the mirrors of the cavity and passes through the laser active medium many times. The resonant cavity thus provides a feedback mechanism, and laser oscillation builds up from the spontaneous emission. Laser threshold occurs when the round-trip gain exceeds the round-trip losses. Losses, including output coupling, are typically a few per cent, and only a few per cent gain is needed for laser threshold.

Since dye lasers can have gain over a wide wavelength band, oscillation can take place anywhere within the gain bandwidth. Without a frequency-selective mechanism, dye lasers usually oscillate at the peak of the gain with typical linewidth of about 50 Å.

By incorporating frequency-selective elements into the cavity, dye lasers can be made to oscillate in a narrow bandwidth and the wavelength of oscillation tuned to anywhere within the gain band. A detailed description of types of cavities and tuning arrangements is given in Sec. IV.

C. Excitation Requirements for Dye Lasers

Dye lasers are pumped optically. The molecules are raised to the excited singlet state by subjecting a volume of dye solution to radiation from a laser or from a high-intensity flashlamp.

We now derive simple expressions for the excitation power density required to obtain optical gain in organic dye solutions. The derivation is made for steady-state conditions, which apply either to continuous operation or to pulsed operation when the length of the pumping pulse is longer than the longest lifetime existing in the system (usually this is τ_T, the triplet lifetime). Let us consider a dye volume having cross-sectional area A and length L. The dye is being excited by optical radiation impinging on the area A and absorbed along the length which is the lasing axis as well. In this longitudinal pumping geometry the pumping source is usually another laser. Neglecting for the moment all losses,

$$\alpha(x) = N_1(x)\sigma_e \tag{10}$$

where x is measured along the laser and pumping axis. For operation very close to threshold the rate equation for $N_1(x)$ is

$$\frac{dN_1(x)}{dt} = N_0 \sigma_{ap} n_p(x) - \frac{N_1(x)}{\tau} \tag{11}$$

where the subscript p refers to the pump wavelength and $n_p(x)$ is the pump photon intensity at the point x. After steady state has been reached, $dN_1/dt = 0$ and

$$\alpha(x) = \sigma_e N_1(x) = \tau N_0 \sigma_e \sigma_{ap} n_p(x) \tag{12}$$

The total exponential gain $G = \int_0^L \alpha(x) \, dx$,

$$G = N_0 \sigma_e \sigma_{ap} \tau \int_0^L n_p(x) \, dx \tag{13}$$

where we have assumed that $N_1/N_0 \ll 1$, hence N_0 is essentially independent of x. The pump photons are being absorbed according to

$$n_p(x) = n_{p0} \exp^{(-\sigma_{ap} N_0 x)} \tag{14}$$

where n_{p0} is the input pump photon intensity. Combining Eqs. (12) and (13) we obtain

$$G = a n_{p0} \sigma_e \tau \tag{15}$$

where $a = 1 - \exp(-\sigma_{ap} N_0 L)$ is the fraction of the pump power absorbed by the dye. Typically, $\sigma_e \approx 10^{-16}$ cm^2, $\tau \approx 5 \times 10^{-9}$ sec. Thus, to achieve a gain of 5%,

$$n_{p0} = 0.05/(10^{-16} \times 5 \times 10^{-9}) \approx 10^{23} \text{ photons/sec-cm}^2$$

This corresponds to a pumping intensity of about 4×10^4 W/cm^2 at 5000 Å. It is clear from Eq. (14) that the shorter the fluorescence lifetime, the higher the required pumping intensity. For a given peak value of σ_e a broad fluorescence band results in a short τ. Thus the very characteristic (broad bandwidth) that makes dye lasers desirable also results in a high pump intensity requirement.

Now let us include the singlet and triplet losses into a more complete expression for the gain. The value of the ground-state molecular density N_0 can be approximated by

$$N_0 \approx N - a n_{p0} / L \tag{16}$$

where N is the total dye molecular density. The rate equation for the lowest triplet state is

$$\frac{dN_T}{dt} = N_1 k_{ST} - \frac{N_1}{\tau_T}$$

(17)

where k_{ST} is the triplet crossover rate, and τ_T is the triplet lifetime. For steady state $(t \gg \tau_T)$,

$$N_T(x) = N_1(x) \, k_{ST}\tau_T$$

(18)

Combining Eqs. (8), (13), (16), and (18), the gain, including the intrinsic losses of the dye, is

$$G = \frac{P_p}{Ah\nu_p} \tau \left[\sigma_e \left(1 - k_{ST}\tau_T \frac{\sigma_T}{\sigma_e} \right) - \sigma_a^* \right] - \left[NL - a \frac{P_p}{Ah\nu_p} \tau \right] \sigma_a$$

(19)

where P_p is the pump power in watts. The above equation has been derived assuming a constant value of the pumped cross section A over the length L, and uniform power density over the area A. More exact calculations, taking into account the Gaussian mode characteristics of the exciting and lasing beams, can be performed. The resulting quantitative differences are usually small, and Eq. (19) remains a very good approximation for most cases of practical interest.

Equation (19) indicates the importance of the triplet losses. The quantity $k_{ST}\tau_T\sigma_T/\sigma_e$ can be of the order of unity even for dyes with very high quantum efficiency if τ_T is long. For a given dye there may exist a value of $k_{ST}\tau_T$ beyond which gain cannot be achieved on a steady-state basis regardless of how strong the excitation. In pulsed dye laser systems triplet losses can be circumvented by using sufficiently fast optical excitation. It is clear from Eq. (17) that during time $t \ll 1/k_{ST}$ the triplet population is negligibly small. Thus if the pump pulse has a risetime $t_2 \ll 1/k_{ST}$, the excited-singlet population can reach threshold value before triplet losses build up.

Singlet losses can be overcome somewhat by sufficiently strong excitation, according to Eq. (19). However, pumping levels required to lift most of the molecules into the excited singlet are very high, and generally singlet losses limit the gain on the short wavelength side of the emission band. For example, in all practical configurations Rhodamine 6G does not lase at the peak of its emission. Choosing dyes for which the fluorescence is shifted considerably away from the absorption band has obvious advantages for laser operation. Very little is known about excited-state absorption in most dyes, but it is evident from Eqs. (8) and (19) that if σ_a^* is large, laser action cannot be achieved regardless of the intensity of the pumping radiation.

Equations (8) and (19) can be used to calculate the gain for a given pump power (or to estimate the pump power needed for laser threshold) if all the quantities are known for a given dye. Conversely, by measuring the gain, fundamental quantities such as σ's, $k_{ST}\tau_T$, etc. can be measured. Some of these applications will be described in Sec. V.

Equation (19) has been derived for the longitudinal pumping configuration. It is easy to show that essentially the same equation is valid for the transverse pumping configuration (pumping axis perpendicular to the laser axis) as well. In that case the area A is still the cross sectional area (normal to the lasing axis) of the pumped region. The area A is now determined by the pumping geometry and by the dye concentration, which fixes the penetration depth at the wavelength of excitation.

It should be noted at this point that there is a limited choice of pumping sources available for a given type of dye laser and for a given excitation source one does not have complete freedom in choosing the dye laser dimensions and other parameters. For example continuous power densities of the order of 10^5 W/cm^2 cannot be realized using conventional light sources. Continuous (CW) dye lasers must therefore be excited by a laser pump. Typical cw pump power available is of the order of 1 W. This power must be focused strongly onto the dye volume. The laws of diffraction then determine the

dimensions of the dye volume, which in turn will largely determine
the type of laser cavity that must be used and the required dye concentration. Practical cw dye laser systems will be described in
Sec. IV.B.3. For flashlamps the maximum available intensity is
limited by the maximum power that can be dissipated in a given volume without destroying the lamp. This fixes the maximum attainable
gain. In addition, the total energy is limited by the amount of
energy that can be supplied to the lamp system. This fixes the
total energy available from the dye laser. For pulsed laser pumping,
on the other hand, the pump light can be strongly focused, and very
high values of gain can be realized. Flashlamps serve to inexpen-
sively bridge the gap between high-power-pulsed-laser excitation
and cw-laser pumping. Practical pulsed dye laser systems are also
detailed in Sec. IV.

D. Dye Laser Tunability and Spectral Condensation

Possibly the most interesting phenomenon associated with organic
dye lasers in terms of physico-chemical processes as well as practical usefulness is the efficient spectral narrowing of dye laser oscillation when a frequency-selective element is inserted into the
optical cavity. A dye laser with a broad-band optical cavity usually
oscillates with a bandwidth of between 50 and 100 $\overset{\circ}{A}$. With the insertion of frequency-selective elements into the optical cavity the
bandwidth of oscillation can be reduced to a small fraction of an
angstrom without appreciable loss in power. Soffer and McFarland
[8] discovered this effect when they substituted a plane diffraction
grating for one of the mirrors of a laser-pumped dye laser. Rotation of the grating resulted in tuning of the dye laser. In Fig. 7
we have illustrated how a plane diffraction grating supplies frequency-selective feedback. Spontaneous emission from the dye laser
medium is scattered off the dispersive grating at an angle dependent
upon the wavelength. Only optical radiation at the desired wavelength of oscillation is returned to the cavity with low loss. The

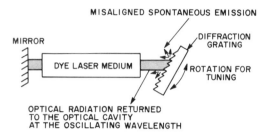

FIG. 7. Frequency selection (tuning) of dye laser with a diffraction grating.

cavity Q is high only in the narrow region of the spectrum where the laser oscillates.

The physical mechanism as to why optical power can be channeled into a small oscillating bandwidth lies in the fact that dye lasers are homogeneously broadened. To explore the meaning of homogeneous broadening as applied to dye lasers let us consider the energy-level diagram in Fig. 8. Normally a dye molecule is excited at some wavelength corresponding to the optical pump, and spontaneous and stimulated emission occur at a range of longer wavelengths according to the Franck-Condon principle. In Fig. 8 we show an energy-level diagram in which molecules have been excited to the first singlet level S_1. Thermalization has taken place, creating a dynamic equilibrium

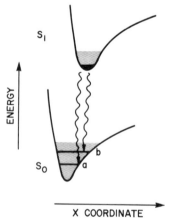

FIG. 8. An energy-level diagram illustrating homogenous broadening in laser dye emission.

in which both the excited state and the ground state are separately
in thermal equilibrium. The levels involved here are a continuum of
rotational and vibrational states. Thermalization means a Boltzman
distribution of occupied levels has been achieved in both the excited
and ground states.

Under lasing conditions with a broad-band optical cavity, stimulated emission can take place from the bottom of the excited singlet
state S_1 to any of the continuum of unoccupied vibrational and rotational levels in the ground state. After the emission of a photon,
rapid thermalization takes place to the bottom of the ground state
on a picosecond time-scale, leaving the terminal level again unoccupied.

Now suppose a frequency-selective feedback is placed in the optical cavity corresponding to the wavelength of a transition to
level a. Laser emission could not take place, for example, to level
b because the cavity Q would be low for that wavelength. The intensity of the optical field in the medium would be strongest at the
oscillating wavelength corresponding to level a. Since the rate of
stimulated emission is proportional to intensity, more molecules are
caused to emit stimulated photons at the wavelength of the intense
field. Since thermalization is so rapid, all molecules in the molecular system have an equal probability of emitting to level a. As
a result the emission to level a draws from the entire excited-state
population, and spectral condensation is therefore very efficient.

IV. DYE LASER DEVICES

In this section we discuss some of the various configurations
in which dye lasers have been constructed. A wide variety of possible laser devices is available to the experimentalist looking for a
dye laser as a laboratory tool.

The primary utility of a dye laser lies with its tunability.
Dye lasers have been constructed which cover the entire spectrum

spanning the near ultraviolet to the near infrared. Coupled closely
with tunability is of course linewidth. The efficient spectral condensation possible with dye lasers allow the reduction of oscillating
bandwidth to be in some cases as small as 4×10^{-6} $\overset{\circ}{A}$. Of course, as
with any laser, optical power and pulse energy are important. De-
pending on the construction, the output of a dye laser can vary from
a few watts continuously to tens of joules in a pulse. Finally, the
device complexity and cost must be considered. This often determined
by whether an expensive laser pump is required or whether an inexpensive flashlamp system will suffice.

A. Flashlamp-Pumped Dye Lasers

Flashlamp pumping of organic dye lasers was first suggested
(based upon results obtained with laser pumping) by Sorokin, Lankard,
Hammond, and Moruzzi [52] in 1967. Since that time many laboratories
have studied flashlamp-excited dye lasers, and they have become commercial products.

The utility of the flashlamp-pumped dye laser lies in its relative simplicity, low cost, high output power, and tunability.
Flashlamp-pumped dye lasers can be constructed which have an output
energy of tens of joules. They have also been able to cover the
entire visible spectrum from near ultraviolet to infrared. One
possible disadvantage is that they are by necessity pulsed devices.

One of the fundamental considerations for flashlamp-pumped dye
lasers is the transient triplet-state problem. As was discussed
previously in Sec. II, the population of triplet levels via intersystem crossing can have a very deleterious effect on dye laser performance. The accumulation of molecules in a long-lived triplet
state has two important effects. First it essentially removes excited molecules from the singlet system, placing them in a state
where they cannot contribute to laser action. Second, molecules in
the triplet state can provide a loss to laser action via triplet-
triplet absorption [see Eq. (8)].

One way to minimize triplet-state effects is to use a flashlamp with a sufficiently fast risetime. Sorokin [52] has pointed out that for a flashlamp pulse with a leading edge that rises linearly with time, reaching a maximum value in a time τ_m, there is a critical pulse risetime where laser action is possible only for $\tau_m < \tau_c$. The critical pump risetime is given by

$$\tau_c = \frac{2\sigma_e}{k_{ST}\sigma_T} \tag{20}$$

This equation can be derived simply by integrating Eq. (17). In the case where nonradiative transitions to the ground state can be neglected, then

$$\tau_c = \frac{2\tau}{1 - \phi} \frac{\sigma_e}{\sigma_T} \tag{21}$$

Typically for xanthene dyes [52] $\phi = 0.9$, $\sigma_e/\sigma_T = 10$, and $\tau = 5 \times 10^{-9}$. Using these values, a critical time becomes $\tau_c \approx 1$ μsec. Thus it appears that at least for dyes in this class the flashlamp risetime should be a few tenths of a microsecond.

It should be noted that with the use of triplet-state quenchers the risetime requirements given by Eq. (21) are relaxed. Snavely and Schafer [53] demonstrated that pulses on the order of milliseconds could be obtained by using oxygen dissolved in methanol as a triplet quencher for Rhodamine 6G. Even longer flashlamp-pumped pulses have been obtained with cyclooctatetraene as a quencher [54]. With the triplet-state restriction removed, the duration of a flashlamp dye laser pulse is determined by the length of the flash and thermally induced index variations which induce loss by scattering light out of the optical cavity.

A typical flashlamp pumping arrangement is shown in Fig. 9. Here we show a linear close-coupled flashlamp arrangement similar to that used for pumping solid-state lasers. The dye is flowed through the dye cuvette to prevent heating by replacing the dye solution between pulses. The optical cavity is formed by a dielectric output mirror and a grating in the Littrow configuration. Rotation

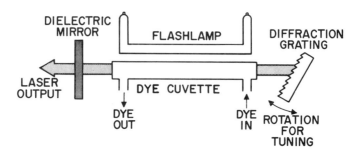

FIG. 9. Typical flashlamp-pumped dye laser in the linear
close-coupled arrangement.

of the grating changes the wavelength of the laser by providing
wavelength-selective feedback to the optical cavity. Other configurations for flashlamps of helical or coaxial design are possible.
Wavelength tuning can also be achieved with a dispersive prism in
the optical cavity.

Flashlamp-pumped dye lasers have been successfully tuned from
3400 to 8000 Å. The number of dyes which have been successfully
flashlamp-pumped is considerably smaller than those that have been
laser-pumped. All flashlamp-pumped laser dyes can be pumped with
laser excitation, but the reverse is not true. Most flashlamp-pumped
dyes in the visible fall into two structural classes, the xanthenes
and the coumarins. In the infrared polymethine dyes have been useful.

Furomoto and Ceccon [55-56] have extended the wavelength range
of flashlamp-pumped dye lasers into the ultraviolet. In order to
pump ultraviolet dyes it was necessary to develop a flashlamp capable
of fast risetime to minimize triplet effects. Also important is that
the flashlamp achieve a high blackbody temperature in order that the
emission from the flashlamp efficiently couple to the short wavelength
absorption band of the organic dye. The results were obtained with
small coaxial xenon flashlamp having a low-inductance, low-capacitance
high-voltage capacitor system. A risetime of 40-50 nsec was achieved
with plasma temperatures in the neighborhood of 65,000°K. With this
system, powerful laser action was obtained at 3410 Å with p-terphenyl

in toluene. Other organic compounds (scintillation detectors) were used to fill in the range from the near UV to the blue region of the spectrum.

A few considerations to bear in mind when using flashlamps for optical pumping is that the radiation is a high-power continuum that can find resonance in the ultraviolet, visible, and infrared. The excited-state configurations of irradiated compounds can be complex. Higher excited states in the singlet manifold can be populated by direct pumping from the ground state. Also, pumping from lower excited states to high excited states in the same manifold is possible.

B. Laser-Pumped Dye Lasers

The high optical power densities achievable from laser sources make them ideal for optically pumping organic dyes. It is not surprising then that the first observation of stimulated emission in an organic dye was observed with laser pumping [2].

The laser-pumped dye laser does have the disadvantage that an expensive coherent optical source is required. However, the spatial coherence of the laser source allows a more precise control over optical pumping geometry than would be possible from an incoherent flashlamp. The continuous dye laser is only possible because a relatively low-power laser beam can be focused into a small volume. The greater power densities achievable from a laser pump provide a means of pumping many dyes which cannot be flashlamp pumped. Also, laser pumps can conveniently provide optical pumping pulses with nanosecond rise times. Laser pumping can be made very efficient, with efficiencies greater than 35% being reported [57].

3. Longitudinal Pumping

The novel scheme of Bradley et al. [58] provides a good example of laser pumping in the longitudinal pumping geometry. In Fig. 10 we show a diagram of this pumping scheme. A Q-switched ruby laser

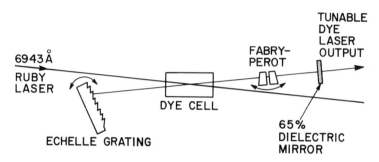

FIG. 10.   Longitudinal pump scheme for a dye laser.

enters the dye laser cavity at a small angle to the cavity axis
(approximately 3°). This avoids the need for special coupling
mirrors. Any ruby light not absorbed in the dye cell passes out
of the cavity. The ruby laser pulse is 20 MW with a 35 nsec pulse
width. The optical cavity for the dye laser is formed by the
echelle grating and the 65% dielectric mirror. The Fabry-Perot
etalon is used for further spectral narrowing.

The results obtained using this configuration are quite im-
pressive. A beam divergence of less than 0.5 mrad and line widths
less than 0.5 Å were obtained without a Fabry-Perot etalon in the
cavity. With an etalon in the cavity, single mode operation was
observed. The line width was measured to be less than 500 MHz.
In fact, the etalon provided a reduction of spectral width of 10^4
at a cost of only a factor of 4 in efficiency. In summary, this
configuration has the advantage of narrow line output, small angle
divergence, and high pulsed output power.

2. Transverse Pumping

Transverse optical pumping is achieved typically by using a
cylindrical lens to focus a laser beam into a narrow strip. The
narrow strip of radiation is then directed onto a dye cell, and
laser action takes place along the strip, transverse to the pumping
radiation.

An example of transverse pumping is shown in Fig. 11. Here we show the rectangular beam from a nitrogen laser being focused by a cylindrical lens onto a cell containing organic dye. The laser cavity is formed by the mirror and the grating.

The N_2 laser is a particularly useful pump for organic dyes. Meyer et al. [59] have obtained laser action spanning the entire visible from 3600 to 7000 Å with several dyes, using the single 3371 Å line as a pump. The N_2 laser used in these experiments produced 10-nsec wide pulses with 100 kW power and was capable of repetition rates of up to 100 pps. One advantage of the transverse pumping scheme is that the dye may be flowed across the pumping region, allowing repetitive pumping. A disadvantage is that

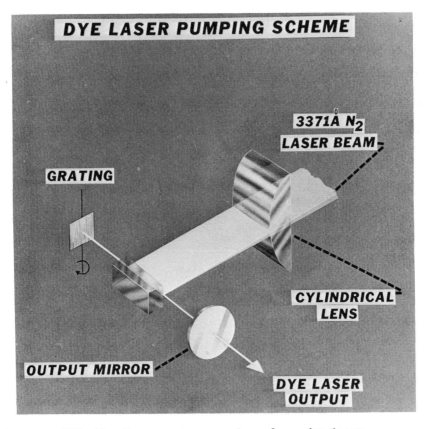

FIG. 11. Transverse pump scheme for a dye laser.

without special care the beam divergence and line width of a transversely pumped laser are greater than for a longitudinally pumped
dye laser [58]. Hansch [60] has demonstrated that a transversely
pumped dye laser can be made to emit in a 0.01 Å bandwidth with an
etalon in the optical cavity. With the use of an additional
passive etalon external to the cavity they have reduced the bandwidth to 7 mHz (or 8 x 10^{-5} Å).

There are a number of useful laser dyes that have very little
absorption at the 3371 Å nitrogen laser wavelength. In order to
excite a sufficient number of molecules to obtain laser action the
concentration of the dye must be increased to the point where
ground-state absorption losses seriously limit the performance. It
is possible to improve the pumping efficiency by using an intermediate dye with larger absorption cross section at 3371 Å whose
fluorescence overlaps the main absorption band of the laser dye.
This is an example of the radiative-transfer quenching described in
Sec. II.C.5. This principle was demonstrated by Moeller, Verber,
and Adelman [61], who used Rhodamine 6G as the intermediate dye to
pump cresyl violet, which has very low absorption at 3371 Å. They
obtained improved laser performance from the mixture compared with
a pure cresyl violet solution.

3. CW Dye Laser

Thus far in our discussion we have limited our attention to
pulsed laser devices. As has been discussed previously, Sorokin
[52] pointed out in early dye laser papers that triple-state accumulation presented a severe problem for continuous dye laser operation. Snavely and Schafer [53] showed that triplet effects could
be overcome by the use of triplet-state quenchers. They produced
a 0.1 msec laser pulse from a flashlamp-pumped solution of Rhodamine 6G in air-saturated methanol. The oxygen dissolved in the
methanol quenched the triplet state sufficiently rapidly to prevent
triplet-triplet absorption from turning the laser off. Pappalardo
et al. [54] showed that even longer pulses of 0.5 msec duration were
observed when cyclooctatetraene, C_8H_8, was used as a triplet

quencher in a solution of Rhodamine 6G dissolved in methanol. The
pulse length in the above experiments was not limited by triplet
effects but by thermal distortions which introduced severe losses
in the optical cavity.

Peterson, Tuccio, and Snavely [10] overcame the problem of
thermal distortion by rapidly flowing the dye solution transversely
to the focus of an argon laser. A population inversion was achieved in the focus of a 1 W argon laser. A small hemispherical cavity
made was very carefully matched with the focused pump beam, and oscillation was observed with 200 mW threshold. In order to improve
the resistance to thermal distortion, water was used as a solvent.
To prevent aggregation of the dye molecules in the water solution,
a detergent was added. A flow of 3-10 m/sec was required to allow
true continuous operation.

A high-efficiency tunable dye laser was reported by Dienes,
Ippen, and Shank [57]. They achieved a 1 W power output with 3.5 W
of pump power. The dye used was Rhodamine 6G dissolved in water
with 4% of soap (Amonyx LO). The laser cavity used in this experiment is shown in Fig. 12. The optical cavity consists of three
mirrors, a prism for tuning, and a Brewster-angle dye cell to minimize cavity losses. The angle θ shown in the figure was chosen to
minimize astigmatism introduced by the off-axis focusing mirror
[62]. The laser was tuned over the range 5600-6550 Å by rotating
mirror M_3.

Hercher and Pike [63] have achieved single-mode operation of a
cw dye laser by placing a Fabry-Perot etalon in addition to a prism

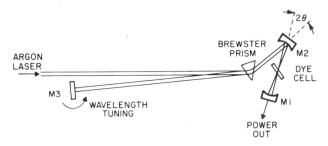

FIG. 12.   CW dye laser pump arrangement.

inside the dye laser cavity. The laser spectrum was reduced to 4×10^{-6} Å with very little loss in output power. This again demonstrates the effectiveness of spectral condensation in homogeneously broadened lasers.

Cw dye laser action has been achieved in other dyes, such as fluorescein [11], Rhodamine B [12], and recently synthesized experimental Rhodamine compounds [12]. At present, tunable cw dye laser power is available from 5200 to 7000 Å [11, 12]. Recently the tuning range of the efficient Rhodamine 6G dye laser has been extended toward shorter wavelengths by using hexafluoroisopropyl alcohol as the solvent [12]. This results in a substantial blue shift of the fluorescence spectrum as discussed in Sec. II.A. A tuning range of 5400-6020 Å was measured for this system. Incidentally, the fluorinated solvent also acts as an antidimerization agent, eliminating the need for the surfactant (soap) used in water solutions.

C. Tunable Distributed-Feedback Dye Laser

Recently it was shown that a new type of "mirrorless" dye laser could be constructed by integrating a periodic structure within a gain medium [64]. One such device is tunable and has a narrow line output. Feedback is obtained by pumping the dye with fringes formed by the interference of two coherent beams.

It is well known that the gain and refractive index of an organic laser dye are altered by the absorption of intense optical radiation. Thus by pumping with two interfering coherent light beams the gain and refractive index may be spatially modulated. This spatial modulation provides a strong frequency-selective coupling between oppositely traveling waves and introduces the feedback necessary for laser oscillation. Strongest coupling occurs when the wavelength in the medium satisfies the Bragg condition, namely, that the period of the spatial modulation must equal one-half the oscillating wavelength in the medium.

A measure of the coupling strength is the coupling constant κ which is given by [65]

$$|\kappa|^2 = \frac{\pi n_1^2}{\lambda} + \frac{\alpha_1^2}{4} \tag{22}$$

where n_1 and α_1 are the amplitudes of the index and the gain modulations, respectively.

In this device it is difficult to tell which of the two terms on the right-hand side of Eq. (22) is dominant. Such a determination would require a detailed knowledge of the organic-molecule matrix elements. However, a simple argument using the Kramers-Kronig relations suggests that both terms could be comparable. It is possible that transient thermal effects and other effects in the solvent might also contribute to the index modulation.

The experimental arrangement is shown in Fig. 13. The optical pump is the second harmonic of a single-mode ruby laser [64]. This 0.347 μ wavelength beam from the KDP doubler is focused, split with a beam splitter into two nearly equal parts, and recombined at the dye cell. The size of the pumped region on the dye cell is approximately 1.6 cm x 0.3 mm; the depth of penetration into the dye is on the order of 100 μ and depends on concentration. The angle θ is varied by changing the mirror positions.

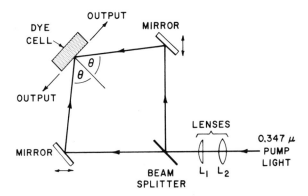

FIG. 13. Experimental arrangement for distributed-feedback pumping.

The wavelength of the dye laser is given by

$$\lambda_L = n_s \lambda_p / \sin \theta \qquad (23)$$

where n_s is the solvent index of refraction at the lasing wave-length λ_L, and λ_p is the pump wavelength. In our experiments we have achieved tuning by varying either θ or n_s.

The result of angle tuning is given in Fig. 14. The dye used was 3×10^{-3} M Rhodamine 6G in ethanol. The points on the figure are experimental; the curve was computed from Eq. (23) accounting for dispersion. The wavelength is quite sensitive to angle tuning with a rate of $d\lambda/d\theta = 80$ Å/deg at 0.6 μ. The range observed for Rhodamine 6G is 640 Å.

We note from Eq. (23) that λ_L is proportional to n_s. The solvent index may be easily varied over a range from 1.33 to 1.55 by proportionate mixing of methanol and benzyl alcohol. A change of 0.01 in n_s changes λ_L by 43 Å.

The laser linewidth was determined to be less than 0.5 Å using a spectrograph. With a Fabry-Perot interferometer having a free spectral range of 50 GHz, the laser spectrum was found to contain

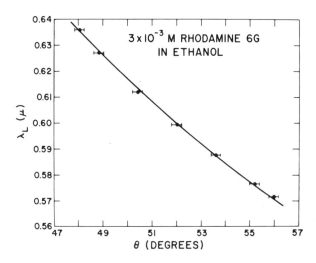

FIG. 14. Distributed-feedback tuning curve.

a number of very narrow lines or modes. With the pump power about
5-14 times threshold, all the lines were contained within a few
tenths of an angstrom. With reduced pump power, apparent single-
mode operation was obtained with a linewidth of less than 0.01 Å,
the resolution limit of the interferometer.

The rather bulky interferometer in Fig. 13 can be replaced by
a prism [66]. The dye solution is placed in contact with a lateral
face of a right-triangular 90° prism. A standing-wave pump field
is produced in the dye solution by interference between the pump
light that passes directly to the inner face in contact with the
dye solution and the pump light internally reflected from the other
lateral face adjacent to the 90° corner of the prism. The output
wavelength is adjusted by changing the angle of incidence between
the pump-laser beam and the prism face.

D. Picosecond-Pulse Mode-Locked Dye Lasers

Because of their wide gain bandwidth, dye lasers are ideally
suited for the production of ultrashort (picosecond) pulses by
mode-locking. A detailed description of the theory and results of
mode locking [67] is beyond the scope of this chapter. Briefly,
however, it can be described as follows: An optical cavity has a
number of resonance frequencies with spacing $\Delta f = c/2L$, where L is
the length of the cavity. Lasers normally operate "multimode,"
i.e., oscillate simultaneously at a number of cavity mode frequencies. If the bandwidth of laser oscillation is $\Delta \nu$, the number of
oscillating modes is $N = \Delta \nu / \Delta f$. Ordinarily the modes have a random
phase relationship, and the power output is constant in time. It
is possible, however, to bring the modes into a particular phase
relationship which results in a repetitive pulsed behavior. The
width of the pulses is inversely proportional to the number of
"locked" modes. It can be easily shown that if all the oscillating
bandwidth $\Delta \nu$ is locked, the pulse width is $\Delta t \approx 1/2\Delta \nu$ and the
repetition rate is $c/2L$.

The best method for achieving mode-locked pulse operation of a laser is to place a saturable absorber in the laser cavity. The saturable absorber is bleached by high-intensity radiation, and it thus acts as a selective gate for short, intense pulses. Almost any dye laser can be mode-locked by this method, making the dye laser a convenient source of tunable picosecond pulses. A typical flash-lamp-pumped mode-locked dye laser would be identical with that shown in Fig. 9 but with a cell containing a suitable saturable absorber (usually another organic dye) placed in the cavity, preferably close to or in contact with the laser mirror. Bradley and coworkers [14, 15] have used such an arrangement to produce continuously tunable pulses of less than 10 psec duration over the wavelength range 5840-7040 $\overset{\circ}{A}$ using a variety of laser and mode-locking dyes. The best results were obtained with Rhodamine 6G as the laser dye and DODCI (diethyloxadicarbocyanine iodide) as the saturable absorber. For this system the pulse duration was as short as 2.5 psec [15].

Recently Ippen, Shank, and Dienes have reported mode locking of the cw Rhodamine 6G dye laser [16], producing a continuous stream of picosecond pulses. The experimental setup is shown in Fig. 15. To

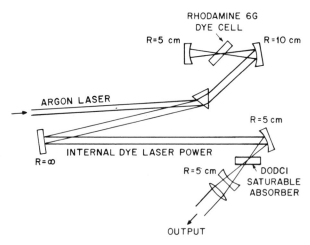

FIG. 15. Experimental scheme for cw picosecond-pulse mode-locked dye laser.

reach sufficiently high intensity for bleaching, the dye laser radiation must be focused into the cell containing the saturable absorber DODCI. This is achieved by adding a second folded mirror arm to the cw dye laser cavity [57]. It is also necessary to have the spot size in the mode-locking dye smaller than in the laser dye. Using this experimental arrangement, pulses as short as 1.5 psec duration were produced and measured.

V. APPLICATIONS OF DYE LASERS

In this section we describe some applications of dye lasers recently reported in the literature. In some of these applications the dye laser is used simply as a tool: a source of tunable co-herent radiation with sufficient power in a narrow bandwidth to excite and probe either atomic or molecular states. For such applications the choice of a given dye laser is made to suit the particular experiment. With the recent explosive growth of dye laser technology a large number of new applications are possible, particularly in the field of spectroscopy. Our purpose here is not to be exhaustive but to show through a few examples the potentials for a great variety of exciting new experiments using tunable dye lasers as light sources.

Another use of dye lasers is in the increased knowledge, and possible further exploration, of the excited states of the lasing dyes themselves. It has been quantitatively shown in Sec. III how the dye laser gain depends on a number of parameters (intersystem crossing rate, excited singlet and triplet absorption cross sections, etc.). While these quantities have often been studied in the past by nonlaser techniques, such measurements were often ex-tremely difficult to perform. The study of the laser properties of the dye allows a direct access to the properties of the higher excited states of these molecules and often allows new and easier

measurements of important parameters. For example, it is clear
from Eq. (19) that the laser gain depends critically on the parameter $k_{ST}\tau_T$. By measuring the laser threshold under a variety of
conditions, Peterson et al. [68] have accurately determined this
important quantity. In Sec. V.C we describe some applications of
well-known photochemical principles to dye lasers and show that
study of dye laser characteristics can lead to increased knowledge
of the photophysics of organic dyes.

A. Gain Spectroscopy

    We discuss here not so much an application of dye lasers but
the use of stimulated emission from organic dyes to study the dyes
themselves. Simple measurements of emission in highly excited
systems with gain are difficult to interpret since emission is
geometry dependent and not directly related to fundamental parameters. Under lasing conditions further complications are introduced by the coupling of molecular emission to the optical-cavity
modes. Direct measurement of gain, on the other hand, provides a
fundamental quantity. In this section we discuss gain spectroscopy
and how it can provide unique information about the excited states
of organic molecules not obtainable from either fluorescence or
absorption spectroscopy.
    A useful and convenient method of measuring gain is based on
measurements of amplified spontaneous emission. This technique
was first demonstrated in high-gain pulsed metal-vapor lasers by
Silfvast and Deech [69] and modified for pulsed dye lasers by Shank,
Dienes, and Silfvast [70]. The experimental arrangement is shown
in Fig. 16. The dye is pumped by 3371 Å light from a pulsed N_2
laser. The gain region is defined by the rectangular pumping beam
which is focused by a 10 cm focal-length cylindrical lens. The
cell windows are canted at a sufficiently large angle to prevent
regeneration. The cell is divided into two sections so that the

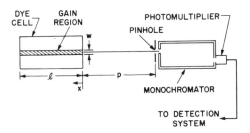

FIG. 16.   Gain spectrometer.

length of the gain region may be varied from ℓ to $\ell/2$ by blocking
off the left half of the pumping beam. After passing through a
250 μ pinhole, the superradiant output is passed through a Jarrel-
Ash 1/4 m monochromator. With the pinhole replacing the entrance
slit the sampled bandwidth is approximately 6 Å. The signal detected by the photomultiplier tube is synchronously detected with a
reference signal of 4 Hz obtained from the pump laser trigger.

For a pinhole coaxial with the gain region and having an area
less than or equal to the cross-sectional area of the gain region,
the light intensity at wavelength λ entering the pinhole is given
by

$$I_\ell(\lambda) = C \int_0^\ell \frac{e^{\alpha_\lambda x}}{(p + x)^2} \, dx \qquad (24)$$

where C is a constant for a given pumping intensity, α_λ is the gain
at wavelength λ, and p and x are defined as shown in Fig. 16. The
quantity α_λ is essentially constant over the 6 Å bandwidth.

For $p \gg \ell$,

$$I_\ell(\lambda) = \frac{C}{\alpha_\lambda p^2} \left[e^{\alpha_\lambda \ell} - 1 \right] \qquad (25)$$

Writing a similar equation for $I_{\ell/2}$ and combining, we get

$$\alpha_\lambda = \frac{2}{\ell} \ln \left(\frac{I_\ell}{I_{\ell/2}} - 1 \right) \qquad (26)$$

Experimentally, gain was measured as a function of wavelength simply by determining the ratio $I_\ell/I_{\ell/2}$ for each desired wavelength. If α_λ varies linearly with pump power, the equation above predicts a linear plot of intensity versus pump power on a semilogarithmic scale. As can be seen in Fig. 17 a linear dependence is observed at low pump powers. At high pump powers deviation from

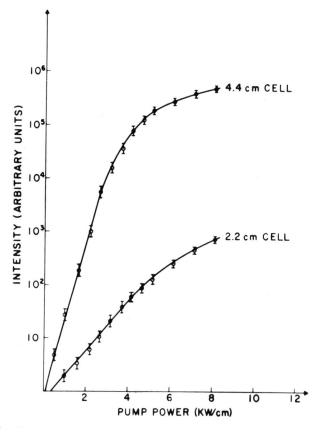

FIG. 17.   Intensity of radiation versus pumping power from 4-methylumbelliferone (4-MU) (4.5 x 10^{-3} M in a 10:1 mixture of ethanol and 1 M HCl). Wavelength is 485 nm. Error bars represent the spread of values from a number of successive measurements.

linearity is seen as the gain saturates. At these powers Eq. (26)
is no longer valid. To determine the gain at high pump powers
shorter cell lengths must be used to prevent saturation.

In Fig. 18 we have plotted gain versus wavelength for
Rhodamine 6G. The circles are experimental and the solid curve is
theoretical. The solid curve was calculated from Eqs. (5)-(7).
The values of the parameters for our system are $N = 9.03 \times 10^{17}$
cm^{-3}, $\tau_f = 5.5 \times 10^{-9}$ sec, and $n = 1.36$. The curves for $\sigma_a(\lambda)$ and
σ_e were taken from Fig. 6. Triplet losses can be neglected in this
measurement since the excitation pulse duration is too short
(10 nsec) to allow a significant accumulation in the triplet state.
A fractional population density in the excited singlet of $N_s/N = 0.0231$

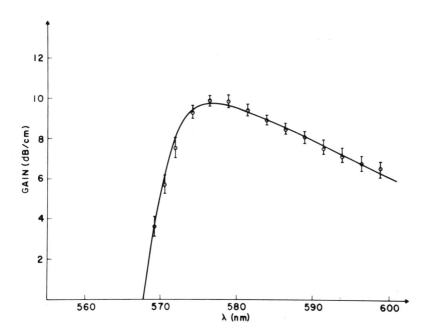

FIG. 18. Gain versus wavelength for Rhodamine 6G at pump
power of 4.2 kW/cm. Circles are experimental points, solid line is
theoretical curve (see text). Fractional excited-singlet population $N_s/N = 0.0231$.

was used to fit the theoretical curve to the experimental data in
Fig. 18. The agreement is seen to be excellent. Thus it appears
that this technique may be a useful method of obtaining excited-
singlet population density.

A more advanced version of this experiment has been assembled
to automatically measure gain as a function of wavelength instead
of taking measurements point by point [71]. A diagram of this
apparatus is shown in Fig. 19. The basis of the system is an in-
cremental digital tape recorder which records the data for processing on a digital computer. Automatic filtering and controls to
advance the monochromator make the system a convenient laboratory
instrument. Various laser dyes, as well as mixtures of dyes, are
presently being investigated with this system. Recently this
apparatus has been used to study the characteristics of the laser
dye 4-methylumbelliferone. These results are described in
Sec. V.C. Typical curves of gain versus wavelength are also shown
in that section.

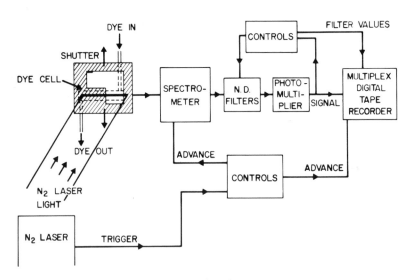

FIG. 19. Automated gain spectrometer.

B. Saturation Spectroscopy with Tunable Dye Lasers

Another interesting application of dye lasers has been demonstrated by Hansch and coworkers [72-74]. In this application,
resonance lines of gases (or other materials) can be studied using
nonlinear spectroscopic techniques [75, 76]. A great advantage of
this technique is that the effects of Doppler broadening which
seriously limit the resolution in ordinary emission or absorption
spectroscopy are eliminated. The now available spectral resolution
and tunability of dye lasers makes it possible to investigate
virtually any optical transition from the near ultraviolet to the
near infrared.

A schematic diagram of a saturation spectroscopy experiment
[73] is shown in Fig. 20. It utilizes a repetitively pulsed tunable dye laser (transversely pumped by a nitrogen laser; see
Sec. IV.B for description). Intracavity frequency-selective
elements reduce the laser bandwidth to 300 MHz (less than 0.004 Å.
An external filter further reduces the bandwidth to 7 MHz (approximately 8 x 10^{-5} Å) [60]. The laser output is divided by a beam
splitter into a stronger saturating beam and a weaker probe beam,
which are sent in opposite directions through the cell containing
the absorbing sample under study. The change of absorption caused
by the saturating beam is detected by the probe beam. A signal is
observed if the laser is close to a resonance frequency so that
both light beams interact simultaneously with the same atoms: those
with close to zero axial velocity. In this manner the actual
resonance frequencies are detected, even though they may be masked
by a broad Doppler profile. The resolution is limited by the laser
bandwidth or by the homogeneous linewidth of the resonances, which-
ever is greater. To increase the sensitivity of the technique the
saturating beam is modulated by a chopper. The resulting amplitude
modulation of the probe beam is detected by a phase-sensitive

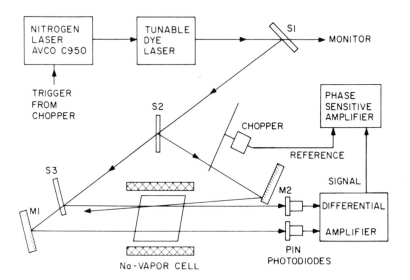

FIG. 20. Experimental arrangement for the use of a dye laser in saturation spectroscopy.

amplifier. The noise due to fluctuation of the laser output is further reduced by using a dummy probe which does not cross the saturated region and by comparing the two probe intensities.

Figure 21 shows some of the results of a study of the red Balmer line of atomic hydrogen. This is of particular interest since the value of the Rydberg constant is based on the wavelength of the Balmer lines. The Doppler width of these lines is extremely large (6 GHz at room temperature), and it is impossible to resolve the fine structure components by conventional spectroscopic techniques. The saturation spectroscopy technique easily resolves the various components of the H_α line and verifies the value of the Lamb shift. The resonance linewidth in the spectrum shown is primarily determined by the natural linewidth and unresolved hyperfine structure. More recent measurements with reduced laser bandwidth revealed even the hyperfine splitting of the metastable 2s state (170 MHz) in the second component from the right.

SPECTRUM OF HYDROGEN

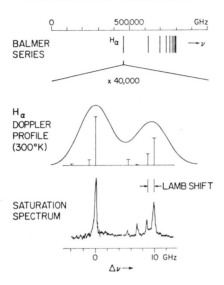

FIG. 21.   Saturation spectroscopy of atomic hydrogen.

In other experiments the saturation spectra of the two sodium
D lines were obtained.   The hyperfine splitting of the ground state
was clearly resolved in both lines and even the splitting of the
excited $^2P_{1/2}$ state was resolved in the D1 line.   Using a probe
beam that was delayed in time and studying the change in intensity
of the saturation signals as a function of time delay, collision
processes perturbing the atomic velocity distributions were also
studied.

C. Modification of the Lasing State

In this section we describe the various ways by which the
emitting state of the dye can be modified to produce new lasing
species.   These processes fall into several types: (a) solvent reorientation, (b) charge-transfer complex formation, (c) dissociation, and (d) protonation.   In many cases the new emitting species

is stable only in the excited state. These cases are particularly
interesting because they provide a means of avoiding the undesirable overlap of the emission spectrum with the absorption spectrum
[77].

1. Solvent Reorientation

As was pointed out in Sec. II.C.3 on solvent effects, many
organic dyes become considerably more polar when electronically
excited. These changes in dipole moment may induce a solvent reorganization which results in an anomalously large Stokes shift of
the fluorescence. This phenomenon has been applied to the case of
p-dimethylamino-p'-nitrostilbene (see structure diagram), which is

$$CH_3 \underset{CH_3}{\overset{}{N}} \!\!-\!\!\bigcirc\!\!-CH=CH-\!\!\bigcirc\!\!-NO_2$$

known to have a much larger dipole moment in the excited state than
in the ground state. Schäfer [6] found that the threshold in this
dye laser system is decreased when benzene/cyclohexane is replaced
as solvent by the more polar ethyl acetate (Fig. 22). An analogous
case was studied by Gronan, Lippert, and Rapp [78] using 2-amino-
7-nitrofluorene (see structure diagram) in 1,2-dichlorobenzene. It

$$O_2N\!\!-\!\!\bigcirc\!\!\bigcirc\!\!-NH_2$$

was found that with increasing dye laser intensity, the laser
spectrum was blue-shifted by about 100 $\overset{\circ}{A}$. The effect was explained
on the basis of solvent orientational relaxation, since one would
expect threshold to be reached sooner with increasing pump power,
and thus the laser emission would be of higher energy (more nearly
from the "Franck-Condon" state).

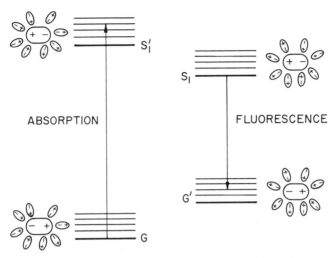

FIG. 22.   Solvent reorientation in excited state.

2. Complex Formation

It is also possible to make use of excited-state complex
formation in creating a larger Stokes shift and consequently a
larger turning range. Schäfer [6] found a significant lowering of
the threshold output and a displacement of the laser wavelength by
about 100 Å for the pyrylium salt shown in the diagram when a small

$$CH_3O \qquad\qquad\qquad OCH_3$$

$$BF_4^-$$

amount of dimethylaniline was added. He suggested that these
events resulted from the formation of a charge-transfer complex in
the excited state such as is known in other systems [41].

A very recent application has been reported by Nakashima et al. [79], who found laser action of the intramolecular exciplex emission of (p-9'-anthryl)-N,N-dimethylaniline in nonpolar and slightly polar solvents. A blue-shift of about 30 Å in a laser spectrum was achieved by increasing the concentration of the dye from 2×10^{-4} M to 1×10^{-2} M, while the spontaneous fluorescence spectrum showed a red shift of about 50 Å with the same concentration increase. These observations were also explained on the basis of laser emission from the nonrelaxed state with respect to solvent reorientation, while the spontaneous fluorescence occurs from a relaxed state.

3. Dissociation and Protonation

a. Studies of the Laser Dye 4-Methylumbelliferone. In this section we describe the results of a recent series of experiments [77, 80-83] on the laser dye 4-methylumbelliferone (4-MU). The behavior of this dye under various solvent environments provides an illustration of the application of a number of photophysical principles described in Sec. II. In particular, ground-state dissociation and excited-state tautomerization takes place for this system. In an initial experiment [77] with 4-MU, laser action from the near UV to the yellow was observed from a single solution of 4-MU which was pumped by a nitrogen laser (see Sec. IV for description). From measurements of the absorption spectrum and the (uncorrected) fluorescence spectrum three different dye forms were identified: the neutral 4-MU, the "basic" form (anion), and the "acid" form which was attributed to an excited-state protonation reaction whose product appeared to have no stable ground state. Some of these results are shown in Fig. 23.

Later experiments [80] on the time-resolved spectroscopy of stimulated fluorescence from the "acid" form provided additional

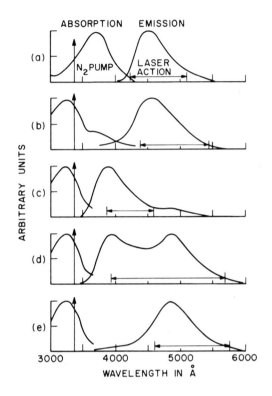

FIG. 23. Absorption and emission spectra of the various forms of 4-MU. The emission spectra were obtained with 3371 Å excitation. The bandwidth of tunable laser action is also shown. (a) Basic form: 4-MU in ethanol with 1 part in 80 of 0.1 M NaOH added; (b) basic and neutral forms: 4-MU in 60% ethanol and 40% water; (c) neutral form: 4-MU in ethanol with 1 part in 400 of 0.037 M HCl added; (d) neutral and acid forms: 4-MU in ethanol with 1 part in 30 of 0.037 M HCl added; (e) acid form: 4-MU in ethanol with 1 part in 10 of 0.37 M HCl added.

evidence that indeed an excited-state reaction was taking place. Evidence was also found that not one but two excited-state reaction products capable of emission were formed.

In a recent investigation of both the spontaneous and laser emission properties of 4-MU [81] the absorption and corrected fluorescence spectra were measured in various solvents, and the laser-pumped gain spectra were determined using the gain spectroscopy apparatus described in Sec. V.A. No less than five different forms, all capable of laser action, were identified. In addition information was obtained about excited-state or triplet absorption that exist for some of the forms. Finally an interesting quenching of one of the excited-state reaction products was observed. These results, which are summarized below, serve as interesting examples of the modification of the lasing state and illustrate the potentialities of gain spectroscopy.

The neutral form (I), which has absorption and emission peaks at 3200 and 3870 Å, respectively, can be isolated in a methanol solvent. In Fig. 24(a) the gain spectrum of such a solution (5×10^{-3} M) is compared with the $E(\lambda)\lambda^4$ spectrum [see Eqs. (5) and (6)]. Considerable losses other than ground-state absorption are indicated. These are thought to be due to either excited-state or triplet absorption.

The anionic form (II), which has absorption and emission peaks at 3640 and 4530 Å, is isolated in basic solvents (for example in methanol with a few per cent of 0.1 NaOH added). This form is due to the ground-state dissociation reaction

$$HO + H_2O \rightleftharpoons {}^-O + H_3O^+ \qquad (27)$$

$$(I) \qquad\qquad (II)$$

In Fig. 24(b) the gain spectrum of such a solution (5×10^{-3} M) is compared to the $E(\lambda)\lambda^4$ spectrum. Here there is good agreement on the short wavelength side but some additional losses are indicated on the long wavelength side. The pump power used to obtain Fig. 24(b) was approximately one-fourth of that used to obtain Fig. 24(a), indicating that form II is a better laser substance than form I, probably because of a higher emission cross section and lower losses.

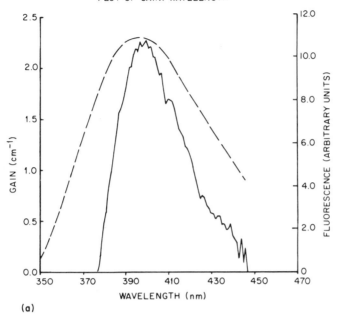

(a)

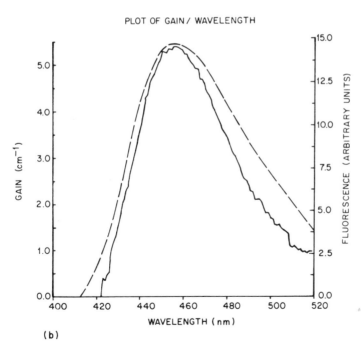

(b)

FIG. 24. Gain spectrum (solid lines) of 4-MU solutions;
dashed line is $E(\lambda)\lambda^4$ spectrum. (a) Neutral form (I); 5 x 10^{-3} M
4-MU, in absolute methanol; pump power: 100%. (b) Anionic form II);
5 x 10^{-3} M 4-MU, 4 x 10^{-3} M NaOH in 96% aqueous methanol.

If water of exactly pH 7.0 is used as a solvent, a mixture of
forms I and II is obtained. A solution of 4-MU in 60:40 methanol-
water was found to contain approximately 40% form II. The gain
spectrum of such a solution (5 x 10^{-3} M) was found to be identical
in shape to that of the pure anionic solution [Fig. 24(b)]. This
indicates that lasing of the neutral form (I) is completely quenched
by the absorption of form II, which overlaps the emission of I. The
peak gain for this water-methanol solution was 4.15 cm^{-1}, approximately 78% of that of the basic solution. From these results it can
be concluded that that portion of the pump power which is absorbed
by form I is transferred to form II. This is an example of the
radiative transfer process described in Sec. II.C.5. The effective
pump power is

$$P_{eff} = P(C_i \phi_I + C_{II}) \qquad (28)$$

where P is the total pump power, C_I and C_{II} are the fractions absorbed by I and II, and ϕ_I is the quantum efficiency of form I.
From the above results we have estimated ϕ_I to be around 64%.

The addition of dilute acid to a 4-MU solution results in a
dramatic 1000 Å shift of the fluorescence peak, while the absorption
remains unchanged. We have suggested earlier [77, 80] that this was
due to an excited-state protonation of the carbonyl group. However,
recent experiments by Nakashima et al. [84], by Yakatan et al. [85],
and by us [81, 83] suggest that the excited-state reaction is a
tautomerization with the emitting species III being the tautomer:

(III)

Our experiments have shown that in very weakly acidic media,
the neutral molecule (I), the anion (II), and the excited-state
tautomer (III) all emit. Figure 25 shows the corrected fluorescence

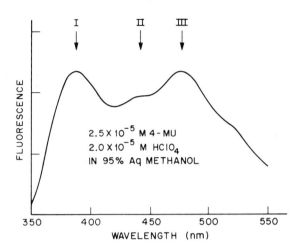

FIG. 25. Corrected fluorescence of a weakly acid 4-MU solution; 2.5×10^{-5} M 4-MU, 2×10^{-5} M $HClO_4$ in 95% aqueous methanol.

spectrum obtained from such a solution (2.0×10^{-5} M $HClO_4$ in methanol containing 5% water). Additional information was obtained with the use of gain spectroscopy. It was found that the shape of the gain spectrum of an acidified solution of 4-MU depended on the pump power: there was relatively less gain at high pump powers in the region 4100 Å, as shown in Fig. 26. The above observations can be explained by the model shown in Fig. 27. One would expect 4-MU to become both a stronger acid (at the phenol group) and a stronger base (at the carbonyl group) in the excited state I^*. Thus the first excited-state reaction in these weakly acidic solvents is a rapid reversible loss of a proton to form the excited anion II^* (this is a unimolecular reaction and would be expected to be independent of hydrogen ion concentration). This rapid step is then followed by a proton transfer from the solvent to the excited anion II^* to form the phototautomer (III^*), which is stable only in the excited state in this solvent environment.

The ground state of the anion in this solution is also "unstable" (no absorption peak due to II could be detected by the use of a Cary 14 absorption spectrophotometer). The reaction $II + H^+ \rightarrow I$ would, however, proceed at a slower rate than $III \rightarrow II$

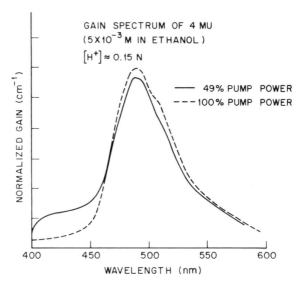

FIG. 26. Gain spectrum of an acidified 4-MU solution;
5 x 10^{-3} M 4-MU.

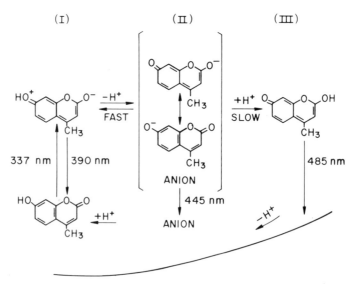

FIG. 27. Suggested model for the reactions of 4-MU in
slightly acid solvents.

(the former reaction is bimolecular, the latter is unimolecular), and at higher pump power there would be a transient accumulation of ground-state anions (II) which absorb beyond 4000 Å and thus would decrease the gain in that region.

Gain spectroscopy has also confirmed the existence of a second excited-state reaction species (IV*) [82]. A fifth form (V), also due to a ground-state reaction has also been identified in concentrated H_2SO_4 solvents. This strong acid form has absorption and emission peaks at 3470 and 4240 Å, respectively, and is also capable of laser action. The existence of this form was first reported by Yakatan et al. [85] and Nakashima et al. [84], who suggested that it is the protonated form of ground-state 4-MU. We have measured the fluorescence and the gain spectra of a solution of 4-MU in a mixture of 92% methanol and 8% concentrated H_2SO_4 and found it contained forms III, IV, and V.

The multiplicity of different forms of 4-MU allows for obtaining a great variety of gain spectra covering the range from 375 nm to above 600 nm. It is in fact possible to "tailor" the shape of a gain spectrum to suit a particular laser requirement by choosing the appropriate solvent. Optical gain over a range as large as 2250 Å has been measured for a single solution of 4-MU [81].

Table 4 is a summary of the various forms and mixtures of 4-MU, together with wavelength range over which gain has been measured with the gain spectroscopy apparatus. Solvent compositions are also given. The concentration of 4-MU was 5 x 10^{-3} M, the percentages for the solvent are by volume.

D. Ultrasensitive Detection of Gain or Loss of Narrow Spectral Lines with a CW Dye Laser

Recently it has been demonstrated that extremely small losses or gains of spectral lines may be detected by inserting samples within the cavity of an oscillating dye laser [86-90]. The influence of either the gain or absorber placed in a broad-band dye

TABLE 4

Summary of Forms and Mixtures of 4-MU

Form[a]	Measured wavelength range of gain (nm)	Solvent
I (neutral)	378-446	Methanol (absolute)
II (anion)	421-545	96% methanol, 4% 0.1 N NaOH (in water)
III* + IV*	450-600	60% methanol, 40% 0.37 N $HClO_4$ (in water)
V	Not measured ~400-480	Concentrated H_2SO_4 or $HClO_4$
I + III* + IV*	380-602	94% methanol, 6% 0.37 N $HClO_4$
I + II + III* + IV*	410-570	80% methanol, 20% water pH6
I + II* + III* + IV*	375-600	94.5% methanol, 5.5% 3.5×10^{-3} N $HClO_4$ (in water)
I + III* + IV* + V	370-580	Methanol + 8% conc. H_2SO_4

[a] Starred items occur in the excited state only.

laser cavity is to cause spectral variations in the laser emission which may be observed with a spectrograph. Since the dye laser oscillating spectrum is extremely sensitive to small frequency-selective perturbations, this constitutes a very effective method for detecting small losses or gains.

In an example involving the detection of small losses, Hansch et al. [89] have shown that it is possible to obtain a 10^5 increase in detection sensitivity over a single-pass absorption method. The tremendous sensitivity of the broad-band laser to small frequency-selective losses is qualitatively understood when we consider the laser gain band to be homogeneously broadened. Then even an infinitesimal increment of losses of modes in one region of the spectrum would cause these modes to be completely quenched and the power to be distributed among the remaining modes. Hansch has shown that the sensitivity of the technique is proportional to the number of oscillating modes, the reason being that in a multimode laser the total intensity for saturation is distributed over many modes, and a sizable decrease of an individual mode intensity is required to compensate for any additional selective loss. The variation in total intensity will be very small, since the gain of the remaining modes is increased, which leads to their greater utilization of the available power.

This technique has been used to study atomic and molecular absorptions in flames [87] and gas samples. Sodium [86] and I_2 [89] have also been studied with this technique.

Klein [90] has shown that the technique can be extended to the study of gain in inverted transitions. In this case the inverted atomic transition provides a small amount of gain in a narrow spectral region. The dye laser broad-band emission is condensed into the small spectral region having gain and appears as a sharp spectral component locked to the frequency of the atomic transition.

This technique has been applied to the study of small gains in a helium-selenium gas discharge [91]. Gain was detected on nine transitions, including five which had not been observed previously as laser transitions. The sensitivity of the technique was such that 0.03% gain was required to produce locking. In Fig. 28 is shown the spectral output of the dye laser and the spontaneous emission of the selenium transition. (Note that the laser line is even narrower than the spontaneous emission.)

VI. CONCLUDING REMARKS

In an area which has shown such rapid progress in the last few years, it is difficult to present a chapter which is up to date by

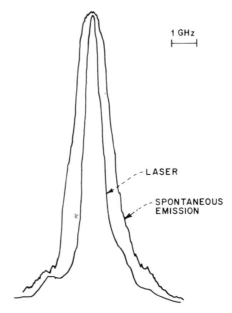

FIG. 28.   Spectral output of a dye laser locked to an atomic transition in a He-Se discharge.   Also plotted is the experimentally measured spontaneous emission.

the time it is published.   To balance this inevitability, we will try to project what will be available to the chemist in the next few years in the field of dye lasers.

There is much to be done in the area of tunable continuous dye lasers.   The wavelength range of continuous dye lasers should be extended into the blue and the near ultraviolet region of the spectrum to provide a continuously tunable source from the near ultraviolet to the near infrared.   The bandwidth of the dye laser probably will be reduced to the kilohertz range in the near future, revolutionizing the field of ultrahigh-resolution spectroscopy.

In the area of tunable short-pulse spectroscopy, continuous trains of picosecond pulses in the 10^3 W range should be available throughout the visible region of the spectrum.   The continuous stable trains of pulses should lead to the development of new and

powerful experimental techniques in the field of picosecond
spectroscopy.

In addition to the above applications in spectroscopy, the
employment of a coherent, high-intensity, wavelength-selective
source will allow detailed study of photochemical reactions where
population of the reactive excited state may be accomplished
directly instead of through the use of energy transfer
(sensitization). Finally, the achievements of gain spectroscopy
in detecting small concentrations of reactive transient species
demonstrate the potential of this technique in studies of the
photophysics of luminescent molecules.

REFERENCES

1. D. L. Stockman, W. R. Mallory, and K. F. Tittel, Proc. IEEE,
   52, 318 (1964).

2. P. P. Sorokin and J. R. Lankard, IBM J. Res. Develop., 10, 162
   (1966).

3. F. P. Schäfer, W. Schmidt, and J. Volze, Appl. Phys. Letters,
   9, 306 (1966).

4. F. P. Schäfer, W. Schmidt, and K. Marth, Phys. Letters, 24A,
   280 (1967).

5. B. B. Snavely, Proc. IEEE, 57, 1374 (1969).

6. F. P. Schäfer, Angew. Chem. (Intern. Ed. Engl.), 9, 9 (1970).

7. M. Bass, T. F. Deutsch, and M. J. Weber, in Lasers, Vol. 3
   (A. K. Levine and A. J. DeMaria, eds.), Dekker, New York,
   1971, Chap. 3.

8. B. H. Soffer and B. B. McFarland, Appl. Phys. Letters, 10,
   266 (1967).

9. P. P. Sorokin and J. R. Lankard, IBM J. Res. Develop., 11,
   148 (1967).

10. O. G. Peterson, S. A. Tuccio, and B. B. Snavely, Appl. Phys.
    Letters, 17, 245 (1970).

11. M. Herscher and H. A. Pike, IEEE J. Quantum Electron., 7, 473 (1971).

12. S. A. Tuccio and F. C. Strome, Appl. Opt., 11, 64 (1972).

13. W. Schmidt and F. O. Schäfer, Phys. Letters, 26A, 558 (1968).

14. D. J. Bradley, A. J. F. Durrant, F. O'Neill, and B. Sutherland, Phys. Letters, 30A, 535 (1969).

15. E. G. Arthurs, D. J. Bradley, and A. G. Roddie, Appl. Phys. Letters, 20, 125 (1972).

16. E. P. Ippen, C. V. Shank, and A. Dienes, Appl. Phys. Letters, 21, 348 (1972).

17. H. Meier, Die Photochemie der Organischen Farbstoffe, Springer-Verlag, Berlin, 1963.

18. C. A. Parker, Photoluminescence of Solutions, Elsevier, Amsterdam, 1968.

19. J. B. Birks, Photophysics of Aromatic Molecules, Wiley (Interscience), New York, 1970.

20. E. P. Ippen, C. V. Shank, and A. Dienes, IEEE J. Quantum Electron., 7, 178 (1971).

21. (a) I. P. Kaminow, L. W. Stulz, E. A. Chandross, and C. A. Pryde, Appl. Opt., 11, 1563 (1972); (b) R. L. Fork and Z. Kaplan, Appl. Phys. Letters, 20, 472 (1972).

22. E. L. Wehry and L. B. Rogers, in Fluorescence and Phosphorescence Analysis (D. M. Hercules, ed.), Wiley (Interscience), New York, 1966, Chap. 3.

23. V. V. Zelinskii, V. P. Kolobkov, and I. I. Reznikova, Dokl. Akad. Nauk SSSR, 121, 315 (1958).

24. D. S. McClure, J. Chem. Phys., 17, 905 (1949).

25. T. Pavlopoulos and M. A. El-Sayed, J. Chem. Phys., 41, 1082 (1964).

26. M. Kasha, Radiat. Res. Suppl., 2, 243 (1960).

27. Y. Hirschberg and F. Gergmann, J. Am. Chem. Soc., 72, 5118 (1950).

28. G. N. Lewis, T. T. Magel, and D. Lipkin, J. Am. Chem. Soc., 62, 2973 (1940).

29. B. L. Van Duuren, Chem. Rev., 63, 325 (1963).

30. E. Lippert, Accts. Chem. Res., 3, 74 (1970).

31. N. S. Bayliss and E. G. McRae, J. Phys. Chem., 58, 1002, 1008 (1954).

32. N. Mataga, Bull. Chem. Soc. Japan, 31, 487 (1958).

33. K. Weber, Z. Physik. Chem. (Leipzig), B15, 18 (1931).

34. T. Förster, Z. Elektrochem., 54, 42 (1950).

35. K. Breitschwerdt, T. Förster, and A. Weller, Naturwiss., 43, 443 (1956).

36. A. Weller, in Progress in Reaction Kinetics, Vol. 1, (G. Porter, ed.), Pergamon, London, 1961, Chap. 7.

37. E. Vander Donckt, in Progress in Reaction Kinetics, Vol. 5 (G. Porter, ed.), Pergamon, London, 1970, Chap. 5.

38. G. Jackson and G. Porter, Proc. Roy. Soc. (London), A260, 13 (1961).

39. T. Förster, Angew. Chem. (Intern. Ed. Engl.), 8, 333 (1969).

40. J. B. Birks, in Progress in Reaction Kinetics, Vol. 5, (G. Porter, ed.), Pergamon, London, 1970, Chap. 4.

41. A. Weller, Pure Appl. Chem., 16, 115 (1968).

42. K. Drexhage, 7th Intern. Conf. Quantum Electronics, Montreal, 1972, paper B5.

43. J. T. Warden and L. Gough, Appl. Phys. Letters, 19, 345 (1971).

44. B. I. Stepanov and A. N. Rubinov, Sov. Phys. Usp., 11, 304 (1968).

45. W. Schmidt, Laser 4, 47 (1970).

46. L. G. S. Brooker, Rev. Mod. Phys., 14, 275 (1942).

47. Y. Miyazoe and M. Maeda, Appl. Phys. Letters, 12, 206 (1968).

48. A. C. G. Mitchell and M. W. Zemansky, Resonance Radiation and Excited Atoms, Cambridge, New York, 1934.

49. B. Lengyel, The Laser, Wiley, New York, 1962.

50. O. G. Peterson, J. P. Webb, and W. C. McColpin, J. Appl. Phys., 42, 1917 (1970).

51. A. Yariv and E. I. Gordon, Proc. IEEE, 51, 4 (1963).

52. P. R. Sorokin, J. R. Lankard, U. L. Moruzzi, and E. C. Hammond, J. Chem. Phys., 48, 4726 (1968).

53. B. B. Snavely and F. O. Schäfer, Phys. Letters, 28A, 728 (1969).

54. R. Pappalardo, H. Samuelson, and A. Lempicki, Appl. Phys. Letters, 16, 267 (1970).

55. H. W. Furumoto and H. L. Ceccon, Appl. Opt., 8, 1613 (1969); J. Appl. Phys., 40, 4204 (1969)

56. H. W. Furumoto and H. L. Ceccon, IEEE J. Quantum Electron., 6, 262 (1970).

57. A. Dienes, E. P. Ippen, and C. V. Shank, IEEE J. Quantum Electron., 8, 388 (1972).

58. D. J. Bradley, A. J. F. Durrant, G. M. Gale, M. Moore, and P. D. Smith, IEEE J. Quantum Electron., 4, 707 (1968).

59. J. A. Myer, C. L. Johnson, E. Kierstead, R. D. Sharma, and I. Itzkan, Appl. Phys. Letters, 16, 3 (1970).

60. T. W. Hansch, Appl. Opt., 11, 895 (1972).

61. C. E. Moeller, C. M. Verber, and A. H. Adelman, Appl. Phys. Letters, 18, 278 (1971).

62. H. W. Kogelnik, E. P. Ippen, A. Dienes, and C. V. Shank IEEE J. Quantum Electron., 8, 373 (1972).

63. M. Hercher and H. A. Pike, Opt. Commun., 3, 346 (1971).

64. C. V. Shank, J. E. Bjorkholm, and H. Kogelnik, Appl. Phys. Letters, 18, 365 (1971).

65. H. Kogelnik, Bell Syst. Tech. J., 48, 2909 (1969).

66. S. Chandra, N. Takeuchi, and S. R. Hartman, Appl. Phys. Letters, 21, 144 (1972).

67. P. W. Smith, Proc. IEEE, 58, 1342 (1970).

68. O. G. Peterson, W. C. McColgin, and J. H. Eberly, Phys. Letters, 29A, 399 (1969).

69. W. T. Silfvast and J. S. Deech, Appl. Phys. Letters, 17, 97 (1970).

70. C. V. Shank, A. Dienes, and W. T. Silfvast, Appl. Phys. Letters, 17, 307 (1970).

71. C. Lingel, R. L. Kohn, C. V. Shank, and A. Dienes, Appl. Optics, 12, 2939 (1973).

72. T. W. Hansch, Proc. 3rd Intern. Conf. Atomic Phys., Boulder, Colo., 1972, to be published.

73. T. W. Hansch, I. S. Shalin, and A. L. Schawlow, Phys. Rev. Letters, 27, 707 (1971).

74. T. W. Hansch, I. S. Shalin, and A. L. Schawlow, Nature, 235 63 (1972).

75. C. V. Shank and S. E. Schwarz, Appl. Phys. Letters, 13, 113 (1968).

76. T. W. Hansch and P. Toschek, IEEE J. Quantum Electron., 5, 61 (1969).

77. C. V. Shank, A. Dienes, A. M. Trozzolo, and J. A. Myer, Appl. Phys. Letters, 16, 405 (1970).

78. B. Gronan, E. Lippert, and W. Rapp, Ber. Bunsenges, Phys. Chem., 76, 432 (1972).

79. N. Nakashima, N. Mataga, C. Yamanaka, R. Ide, and S. Misumi, Chem. Phys. Letters, 18, 386 (1973).

80. A. Dienes, C. V. Shank, and A. M. Trozzolo, Appl. Phys. Letters, 17, 189 (1970).

81. A. Dienes, C. V. Shank, and R. L. Kohn, IEEE J. Quantum Electron., 9, 833 (1973).

82. R. L. Kohn, C. V. Shank, and A. Dienes, to be published.

83. A. M. Trozzolo, A. Dienes, and C. V. Shank, J. Am. Chem. Soc., 96 (1974).

84. M. Nakashima, J. A. Sousa, and R. C. Clapp, Nature, 235, 16 (1972).

85. G. J. Yakatan, R. J. Juneau, and S. G. Schulman, Anal. Chem., 44, 1044 (1972).

86. N. C. Peterson, M. J. Kurlyo, W. Braun, A. M. Bass, and R. A. Keller, J. Opt. Soc. Am., 61, 746 (1971).

87. R. J. Thrash, H. von Weyssenhoff, and J. S. Shirk, J. Chem. Phys., 55, 4659 (1971).

88. R. A. Keller, E. F. Zalewski, and N. C. Peterson, J. Opt. Soc. Am., 62, 319 (1972).

89. T. W. Hansch, A. Schawlow, and P. Toschek, IEEE J. Quantum Electron., 8, 802 (1972).

90. M. Klein, Opt. Commun., 5, 114 (1972).

91. M. Klein, C. V. Shank, and A. Dienes, Opt. Commun., 7, 178 (1973).

AUTHOR INDEX

A

Adelman, A. H., 188
Alfano, R. R., 122, 133
Alger, R. S., 64
Antheunis, D. A., 3, 4, 38, 40
Armstrong, J. A., 123
Arthurs, E. G., 104
Atkins, P. W., 74, 75, 87, 89, 90, 91, 92
Auston, D. H., 104, 115
Avery, E. C., 74, 87

B

Bass, M., 167, 150, 213
Bassetti, F., 2
Bayliss, N. S., 161
Beckett, A., 90
Bennett, T. J., 64, 71, 74, 86, 87, 90
Bernhein, R. A., 2, 8
Birks, J. B., 153, 165
Bjorkholm, J. E., 190, 191
Bloembergen, N., 121, 139
Bolton, J. R., 64, 67, 70, 71, 77, 78, 80, 81, 82, 91, 93, 94, 95
Borg, D. C., 64, 67, 78
Boyd, G. D., 107
Bradford, J. N., 104, 140
Bradley, D. J., 99, 104, 115, 145, 152, 185, 188, 194
Braun, W., 213
Breit, G., 2
Breitschwerdt, K., 163
Brienza, M. J., 121
Brooker, L. G. S., 169
Brossel, J., 2
Burnham, D. C., 112

C

Calvert, J. G., 13, 68
Carman, R. L., 121, 139
Ceccon, H. L., 184
Chan, I. Y., 3
Chandra, S., 193
Chandross, E. A., 157
Cho, D. H., 91, 92
Chen, C. R., 37, 46
Chuang, T. J., 124
Clapp, R. C., 210, 213
Colles, M. J., 121, 139
Collins, R., 2, 8, 15

D

Dahlstrom, L., 115
Deech, J. S., 196
de Groot, M. S., 3, 8
DeMaria, A. J., 99, 121
Dennison, E. W., 145
Dienes, A., 104, 145, 152, 157, 185, 189, 194, 195, 200, 204, 205, 206, 207, 210, 213, 215
Doetschman, 71, 72
Drent, E., 132
Drexhage, K. H., 124 166
Duguay, M. A., 99, 119, 134
Durrant, A. J. F., 185, 188
Dutton, P. L., 94

E

Eberly, J. H., 196
Eckardt, R. C., 104, 140
Eisenthal, K. B., 111, 124
Ellet, A., 2
El-Sayed, M. A., 3, 4, 5, 8, 37, 39, 40, 46, 47, 51, 52, 158

223

SUBJECT INDEX

A

Acid-base properties of excited states, 164

Acid-base reactions, in excited state, 206-208

Acridine, 164

Acridine red, 121, 130

2-Aminoanthracene, 164

Amplitude modulation in PMDR, 47-51

(p-9'-Anthryl)-N,N'dimethyl-aniline, 206

Autocollimator, 110

Azulene, 132

B

Bacteriochlorophyll, 93

Benzophenone anion, 90, 92

Benzoquinone, 93

Biradical complexes, 93

Birefringence, 135

Bragg condition, 190

Brewster angle, 100

Broadening line 180

Bubble formation in liquid N_2, 18

C

Carbon disulfide, 134-137

CAT-FPESR interface, 83

CAT techniques in FPESR, 76, 83

Cavity, 100

Cavity configurations for FPESR, 70-75

Charge transfer complexes, 203, 205-206

Chloroanil anion, 81, 91

Chlorophyll, 91

Chlorophyl cation, 93

Chloroplasts, green plant, 94

Chromatium D, 93, 94

Circuit for fast photodiode, 114

Coumarin, 168, 184

p-Cresol, 164

Crocein scarlet, 130

Crossed beams light gate, 135

CW dye lasers, 194

Cyanine, 168

1,3-Cyclooctadiene, 167

Cyclooctatetraene, 167, 188

D

Decay techniques in PMDR, 38-41

Detection in PMDR, 19-29

2,3-Dichloroquinoxaline, 27, 35, 37, 54

Diethylaniline, 165

1,1'-Diethyl-2,2'-dicarbocyanine iodide, 109

Distributed-feedback dye laser, 190-193

Doppler broadening, 201, 203

Dye bleaching, 125

Dye fluorescence, 157-167

Dye lasers in FPESR, 94

Dye spectroscopy, 153-157

E

Eastman Dye A9860, 108, 127-130

Eastman Dye A9740, 108

Echelon, 131

Effect of crystal sites, 24

Efficiency of harmonic generation, 121

Electron transfer, 165

Energy transfer, 166

ESR detection of transient free radicals, 63

ESR spectra, overlapping spectra, 66

227